BIOLOGY
TOPICAL

SHINGLEE PUBLISHERS PTE LTD
120 Hillview Avenue #05-06/07
Kewalram Hillview Singapore 669594
Tel: 6760 1388 Fax: 6762 5684
e-mail: info@shinglee.com.sg
http://www.shinglee.com.sg

Printed and Published by Shing Lee Publishers Pte Ltd for the Singapore Examinations and Assessment Board under licence from the Chancellor, Masters and Scholars of the University of Cambridge (acting through the University of Cambridge International Examinations Syndicate) either solely or jointly with the Government of Singapore (acting through the Ministry of Education, Singapore).

First Published 2010
Revised Edition 2011
Revised Edition 2012
Revised Edition 2013
Revised Edition 2014
Revised Edition 2015
Reprinted 2015

ISBN 978 981 237 983 2

Printed in Singapore

PREFACE

O Level Biology (Topical) 2005-2014 has been written in accordance with the latest syllabus (5158). Questions from past examination papers are compiled according to topics for effective practice. Within each topic, they are separated into Multiple Choice Questions, Structured Questions and Free Response Questions. The number of marks for each question is included. The quotation (e.g. N12/1/1) after each question indicates the year, paper number and question number.

Comprehensive revision notes, hints and model solutions are included in a separate book as an additional resource.

We believe this set of books will be of great help to teachers teaching the subject and students preparing for their O Level Biology examination.

CONTENTS

TOPIC 1
Cell Structure and Organisation

MULTIPLE CHOICE QUESTIONS . . . .

1. The diagram shows the structure of a typical animal cell as seen using an electron microscope.

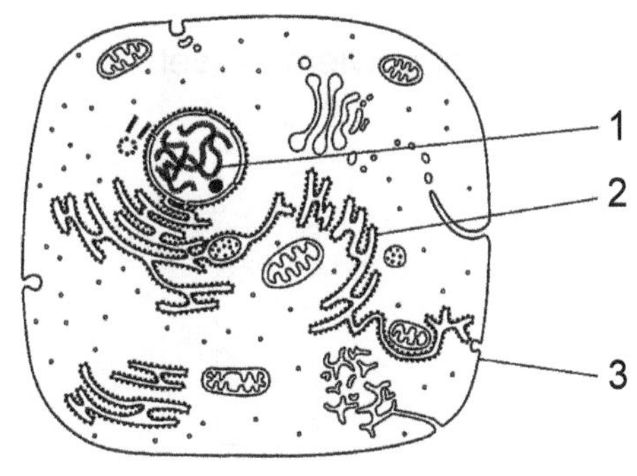

Which cell components are needed to synthesise and build proteins within the cell?

A 1 only
B 1 and 2 only
C 2 and 3 only
D 1, 2 and 3

(B)
[N12/1/1]

2. Which structures are found in a human sperm cell?

	diploid nucleus	haploid nucleus	mitochondria	nuclear membrane
A	✓	×	✓	✓
B	✓	×	×	×
C	×	✓	✓	✓
D	×	✓	×	×

key
✓ = structure present
× = structure absent

( )
[N12/1/2]

3. When viewed through an electron microscope, which structure is surrounded by a double membrane?

A endoplasmic reticulum
B Golgi body
C mitochondrion
D ribosome

( )
[N11/1/1]

4. Which row describes the features of a mature human red blood cell?

	nucleus	large surface area of cell membrane	rigid shape
A	✓	✓	✗
B	✓	✗	✓
C	✗	✓	✓
D	✗	✓	✗

key
✓ = feature present
✗ = feature absent

( )
[N11/1/2]

5. The diagram shows part of a xylem vessel.

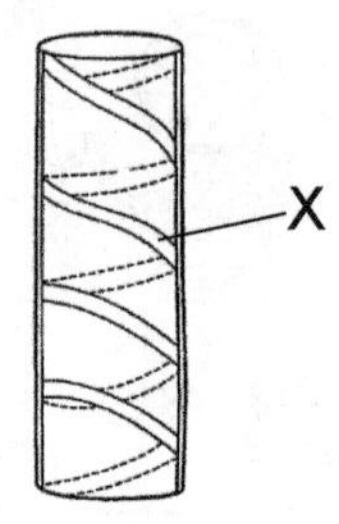

What is the function of the spiral structure X?

A absorption
B photosynthesis
C support
D transport

(C)
[N10/1/1]

6. Which features of a red blood cell would make it an efficient transporter of oxygen to all cells of the body?

	biconcave disc shape to hold extra oxygen molecules	lack of nucleus increases the volume available for oxygen carrying molecules	low surface area to volume ratio to increase diffusion rate of oxygen
A	✓	✓	✗
B	✓	✗	✓
C	✗	✓	✗
D	✗	✗	✓

key
✓ = correct
✗ = incorrect

(A)
[N10/1/2]

7. The diagrams show two cells as seen using a light microscope.

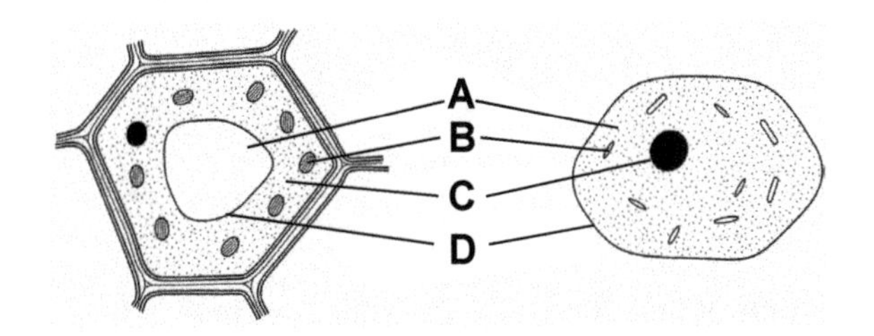

Which label is correct for both cells?

A cell sap
B chloroplast
C cytoplasm
D membrane

( )

[N09/1/1]

8. The diagram shows an animal cell that has some organelles and membrane systems labelled.

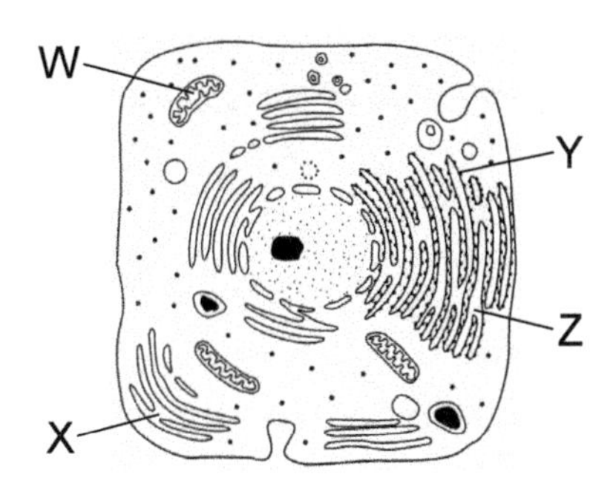

Their functions are:

1 aerobic respiration
2 formation of polypeptides
3 transport of proteins
4 synthesis of fats

Which row in the table gives the correct match between label and function?

	W	X	Y	Z
A	1	2	3	4
B	1	4	2	3
C	2	1	4	3
D	4	3	2	1

( )

[N09/1/2]

9. The diagram shows four types of cell, not drawn to scale. Which cell does **not** contain cytoplasm?

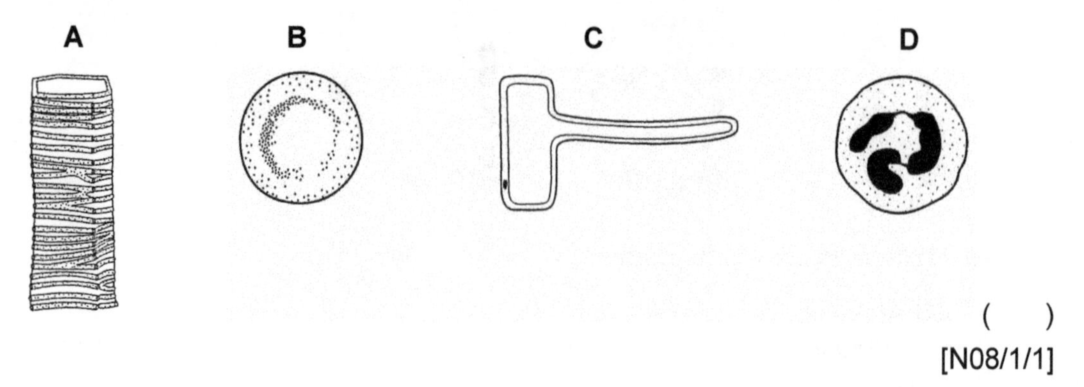

( )

[N08/1/1]

10. The diagram shows an animal cell.

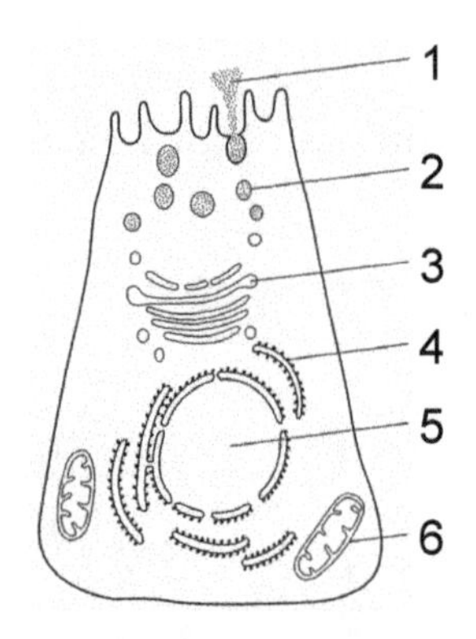

What is the functional relationship between the labelled structures?

A 1 is synthesised by 4.
B 2 develops into 6.
C 3 contains products synthesised by 5.
D 4 controls the contents of 5.

( )

[N08/1/2]

11. Which line in the table correctly identifies these body components?

1 brain, spinal cord and nerves
2 blood
3 neurone
4 stomach

	cell	tissue	organ	organ system
A	2	3	1	4
B	2	4	3	1
C	3	2	1	4
D	3	2	4	1

( )

[N07/1/1]

12. The table shows some characteristics of four types of cell.

Which cell could be a root hair cell?

	nucleus	chloroplast
A	✓	✓
B	✓	×
C	×	✓
D	×	×

key
✓ = present
× = absent

( )
[N06/1/1]

13. The diagram shows a plant cell.

Which labelled structure controls the passage of substances into and out of the cell?

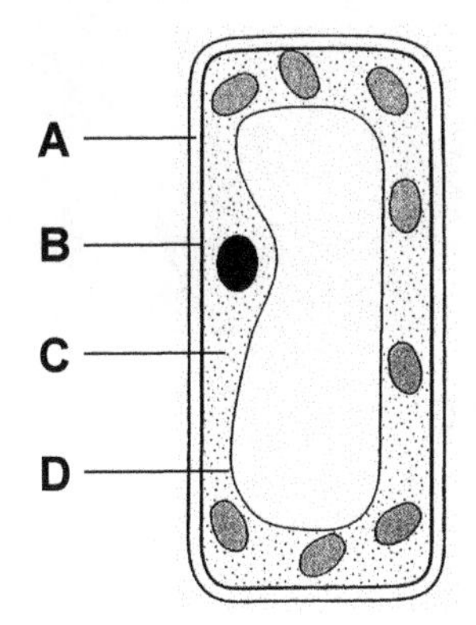

( )
[N05/1/1]

STRUCTURED QUESTIONS . . . .

1. The figure shows an electron micrograph image of part of a cell from the brain of a rat.

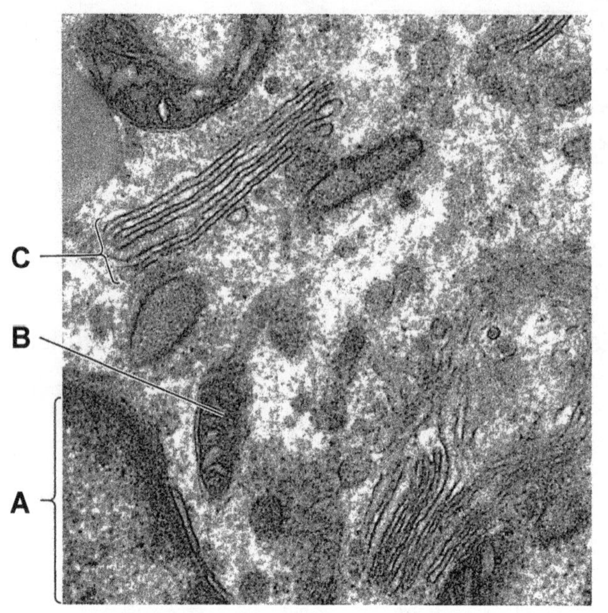

Identify the parts labelled **A**, **B** and **C**.

A...

B...

C...[3]

[N12/2/7(a)]

2. The figure shows the detailed structure of an animal cell.

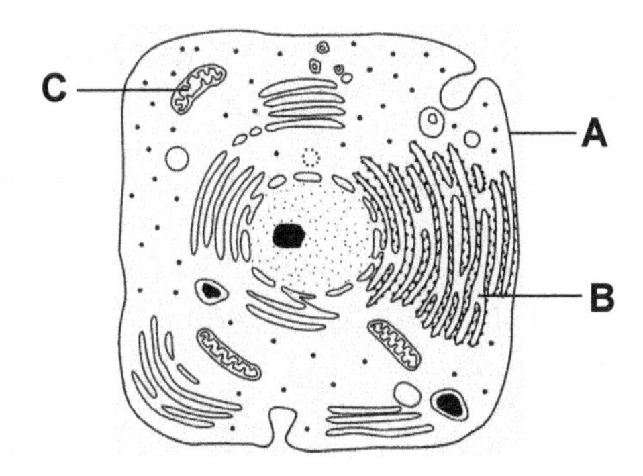

(a) (i) Identify the structures labelled **A** and **B**.

A ....

....

B ....

.... [2]

(ii) Identify structure **C** and name the process that takes place in this structure.

name ....

....

process ....

.... [2]

(b) State two structures that are found in plant cells but not in the cell shown in the figure.

1. ....

....

2. ....

....[2]

[N08/2/1]

FREE RESPONSE QUESTIONS. . . . .

1. **(a)** State the similarities and differences between the structure of plant and animal cells as seen under a light microscope. [4]

   (b) Name two structures in a cell visible only under an electron microscope and state the function of each one. [2]

   (c) Using a **named** plant structure as an example, describe the relationship between cells, tissues and organs. [4]

[N10/2/10 Either]

Movement of Substances

MULTIPLE CHOICE QUESTIONS . . . .

1. The diagram shows a root hair cell, Z, and two root cells, X and Y.

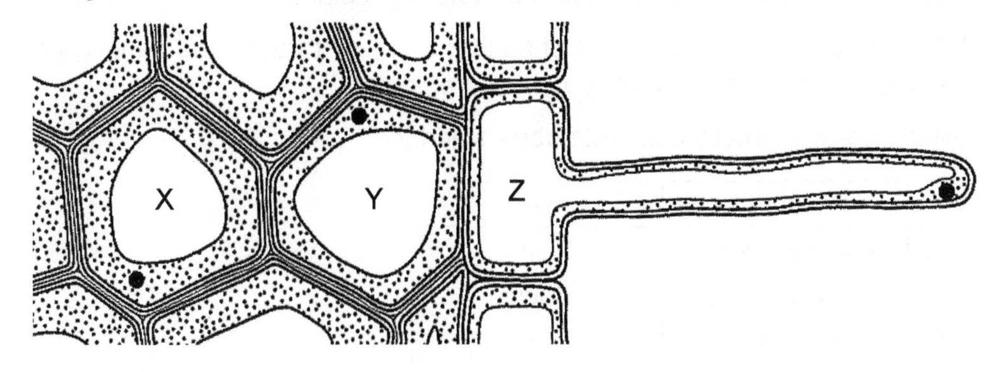

How do magnesium ions and water molecules move from Z to Y and Y to X?

	movement of magnesium ions		movement of water molecules	
	Z to Y	Y to X	Z to Y	Y to X
A	active transport	active transport	active transport	active transport
B	active transport	diffusion	active transport	osmosis
C	diffusion	active transport	osmosis	active transport
D	diffusion	diffusion	osmosis	osmosis

( )

[N12/1/3]

2. What is an example of active transport?

A movement of ions up the xylem
B movement of water into root hairs
C uptake of glucose by cells of villi
D uptake of oxygen by red blood cells

( )

[N11/1/3]
[N07/1/2]

3. A plant cell is immersed in water.

Why does water move into the cell?

A The cell membrane is fully permeable.
B The cell wall is partially permeable.
C Water moves down a water potential gradient.
D Water is actively transported into the cell.

( )

[N10/1/3]

4. The sentence describes the uptake of water by a plant.

Water moves into the root hairs of a plant by ...1..., through the partially permeable cell membrane, ...2... a water potential gradient.

Which words correctly complete gaps 1 and 2?

	1	2
A	active transport	up
B	active transport	down
C	osmosis	up
D	osmosis	down

( )

[N09/1/3]

5. The diagrams show a cell in solution X and the same cell after it had been in solution Y for 20 minutes.

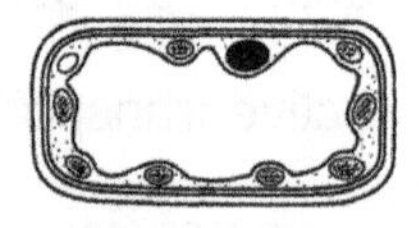
solution X

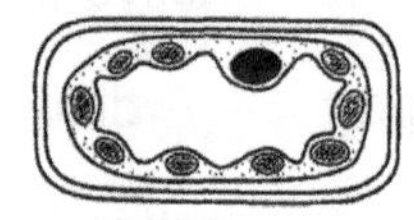
solution Y

What caused the change and what was the relative water potential of solution Y to solution X?

	change caused by water diffusing	water potential of solution Y to solution X
A	in	higher
B	in	lower
C	out	higher
D	out	lower

( )

[N09/1/11]

6. The table shows the concentration of a substance inside and outside four different cells.

Which cell would need the most energy to absorb the substance by active transport?

cell	concentration (arbitrary units) inside cell	concentration (arbitrary units) outside cell
A	3	6
B	3	9
C	6	3
D	9	3

( )

[N08/1/3]

7. The apparatus was set up as shown in the diagram.

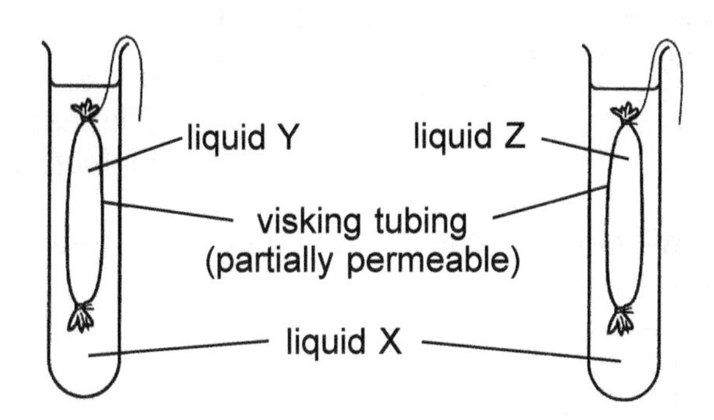

After 30 minutes had elapsed, the partially permeable tubing containing liquid Y had collapsed while the tubing containing liquid Z was firm.

Which could be a correct description of the liquids at the **start** of the experiment?

	liquid X	liquid Y	liquid Z
A	10% sucrose solution	water	25% sucrose solution
B	25% sucrose solution	10% sucrose solution	water
C	water	25% sucrose solution	10% sucrose solution
D	10% sucrose solution	25% sucrose solution	water

( )

[N07/1/3]

8. Which processes can take place in a root hair cell when oxygen is **not** available?

A active transport only
B diffusion only
C active transport and osmosis
D diffusion and osmosis

( )

[N06/1/2]

9. The diagram shows an apparatus used to investigate osmosis.

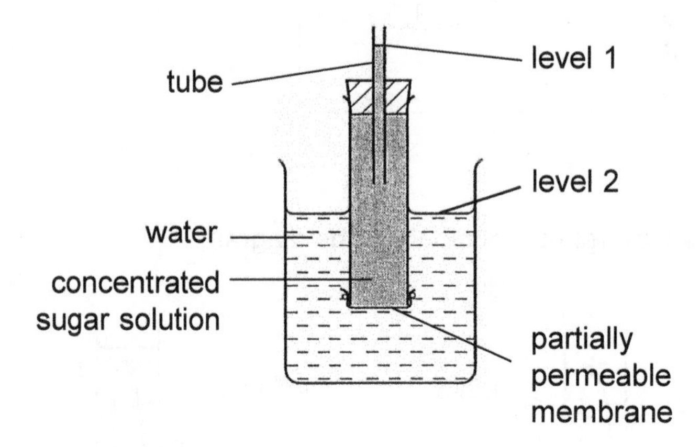

Which molecules will move across the partially permeable membrane and which changes in levels will occur?

	molecules	level 1	level 2
A	sugar	fall	rise
B	water	fall	rise
C	sugar	rise	fall
D	water	rise	fall

( )

[N06/1/3]

10. The diagram shows a root hair, surrounded by a dilute solution of mineral ions.

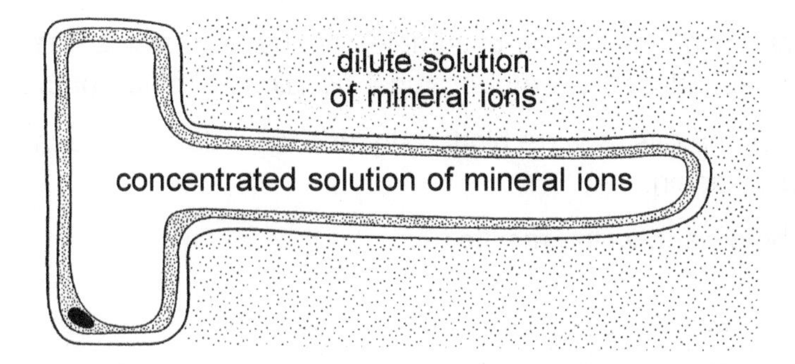

Which statement is correct?

A Water molecules move into the root hair because their concentration is lower inside.

B Water molecules move into the root hair because their concentration is lower outside.

C Water molecules move out of the root hair because their concentration is lower inside.

D Water molecules move out of the root hair because their concentration is lower outside.

( )

[N05/1/2]

11. Which process can involve active transport?

A carbon dioxide intake through stomata

B mineral ion intake through root hairs

C mineral ion transport through xylem vessels

D water leaving mesophyll cells

( )

[N05/1/4]

STRUCTURED QUESTIONS . . . .

1. In an investigation, the volume of samples of 20 dried raisins was measured. Each sample was then placed in water or sugar solutions of different concentrations.

After 12 hours the raisins were blotted dry and the volume of each sample of raisins was measured again.

The figure below shows the results.

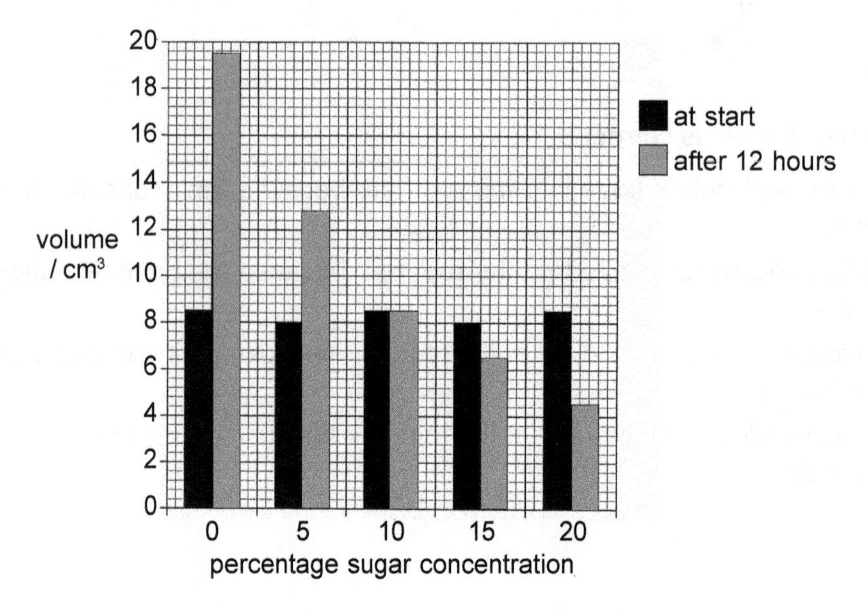

(a) Explain why samples of 20 raisins were used rather than a sample of one raisin.

....

....[1]

(b) Calculate the percentage increase in the volume of the sample of raisins in 5% sugar solution.

....[2]

(c) Explain the results in the 10% sugar solution.

....

....

....

....[2]

[N09/2/7]

FREE RESPONSE QUESTIONS. . . . .

1. **(a)** Define the terms:

(i) *diffusion* [2]

(ii) *osmosis.* [3]

(b) Transdermal patches are adhesive patches which are stuck on the skin.
The patches contain small amounts of medicines.
The inner protective layer is peeled off and the patch is stuck on the skin.
The figure shows a section of a transdermal patch with the protective layer being peeled off.

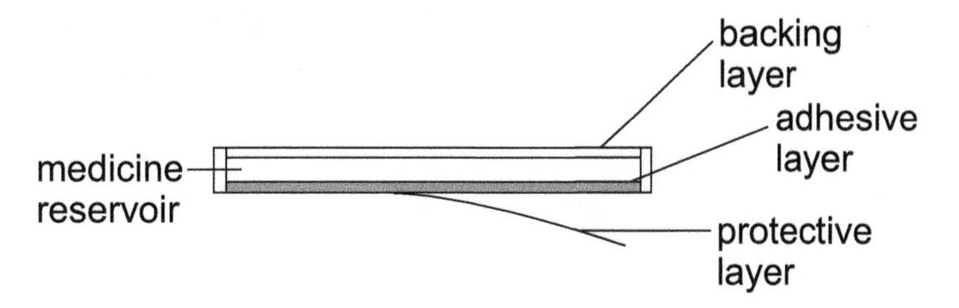

(i) Suggest how the medicine in the patch reaches the blood stream. [2]

(ii) Suggest two advantages of giving medicines using a patch rather than taking medicines through the mouth. [2]

(iii) Suggest **one** disadvantage of giving medicines using a patch. [1]

[N11/2/9]

2. An experiment was carried out to investigate the effect of different concentrations of sugar solution on potato tissue.

Fifty cubes of potato tissue of the same dimensions were cut and weighed.

Ten cubes were placed in pure water and ten placed in each of four different concentrations of sugar solutions.

The cubes were left for one hour.

They were then removed from the solutions, dried carefully with blotting paper and reweighed.

The table shows the results.

concentration of sugar solution g / 100 cm^3	mean initial mass / g	mean final mass / g	mean change in mass / g	mean percentage change in mass (2 s.f.)
0	2.23	2.42	+ 0.19	+ 8.5
5	2.31	2.47	+ 0.16	+ 6.9
10	2.18	2.30	+ 0.12	+ 5.5
20	2.27	2.15	– 0.12	– 5.3
40	2.39	2.20	– 0.19	– 7.9

(a) Plot the mean percentage change in mass against the concentration of sugar solution. [4]

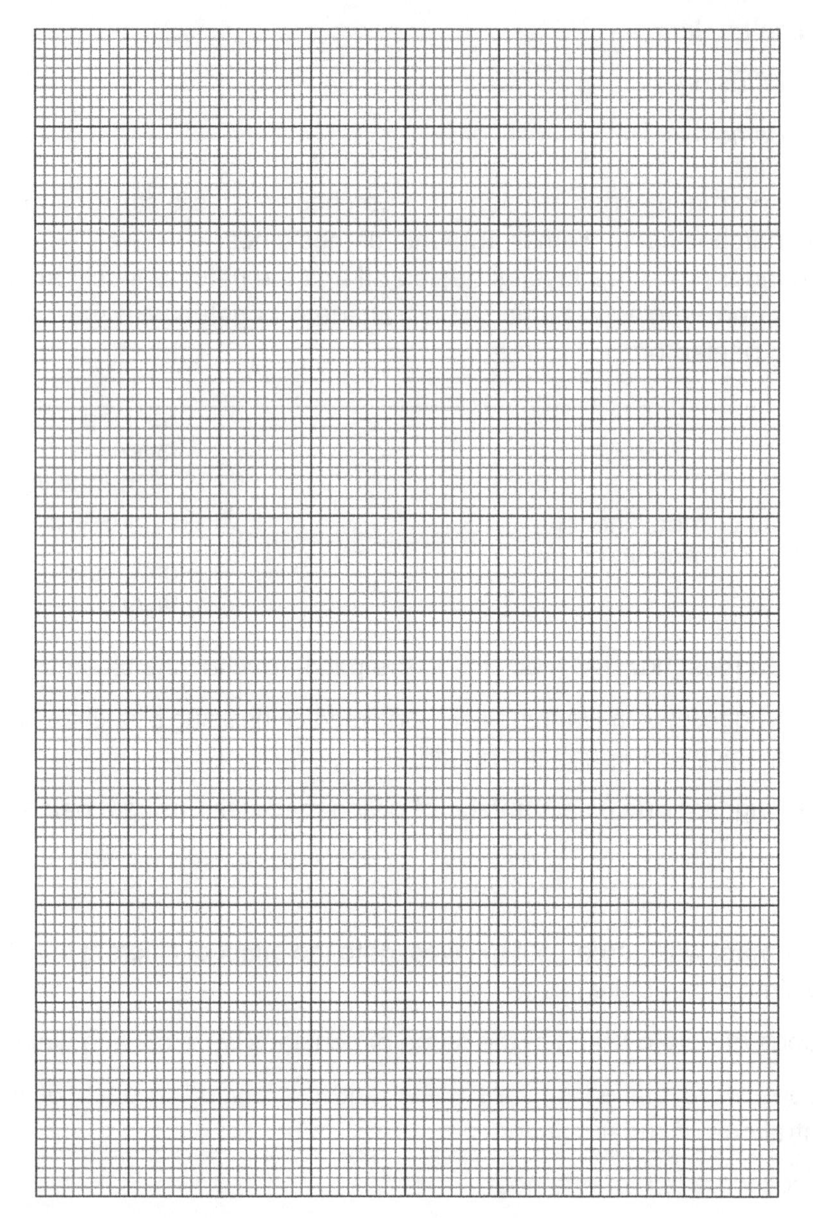

(b) Use your graph to find the concentration of sugar solution at which there was no change in mass. [1]

(c) How is this concentration related to the concentration of the cell sap of potato cells? [1]

(d) From the results of the experiment, state what can be deduced about the cell membrane of the potato cells. [2]

(e) Explain why the potato cubes in 5 g/100 cm^3 sugar solution gained mass. [4]

[N08/2/9]

TOPIC 3
Biological Molecules

MULTIPLE CHOICE QUESTIONS . . . .

1. What happens when an enzyme catalyses a reaction?

A The activation energy is raised.
B The enzyme molecule cannot be used again.
C The products are different in the presence of the enzyme.
D The speed of the reaction changes. ( )

[N12/1/4]

2. Which property of water is important for chemical reactions to take place inside cells?

A It acts as a solvent.
B It has high specific heat capacity.
C It is transparent.
D There is cohesion between its molecules. ( )

[N12/1/5]

3. The table below shows the results of tests carried out on a drink.

test	result
biuret	clear blue colour
Benedict's	orange colour
ethanol emulsion	white emulsion formed
iodine	yellow colour

What does the drink contain?

A fat and reducing sugar only
B protein and starch only
C fat, protein and reducing sugar only
D fat, protein, reducing sugar and starch ( )

[N12/1/6]

4. The diagram shows some uses of water.

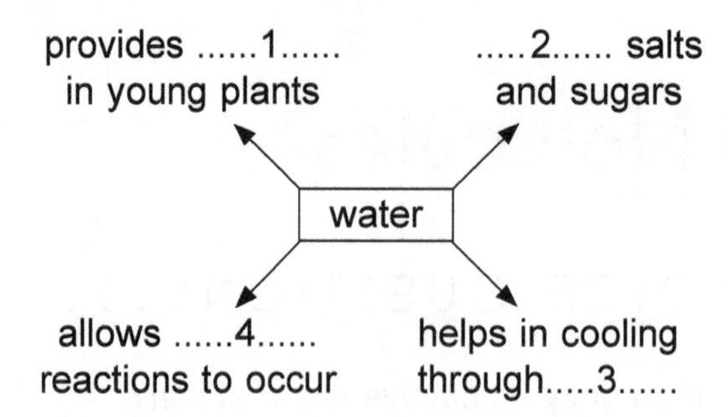

Which words correctly complete gaps 1, 2, 3 and 4?

	1	2	3	4
A	chemical	evaporation	transports	support
B	evaporation	support	chemical	transports
C	support	transports	evaporation	chemical
D	transports	chemical	support	evaporation

( )

[N11/1/4]

5. Milk produces a brick red precipitate when heated with Benedict's solution.
A purple colour develops when the biuret test is used on milk.
Using these results **only**, which nutrients does milk contain?

A fat and protein
B fat and starch
C reducing sugar and protein
D reducing sugar and starch

( )

[N11/1/5]

6. The graph shows the energy changes in a chemical reaction.

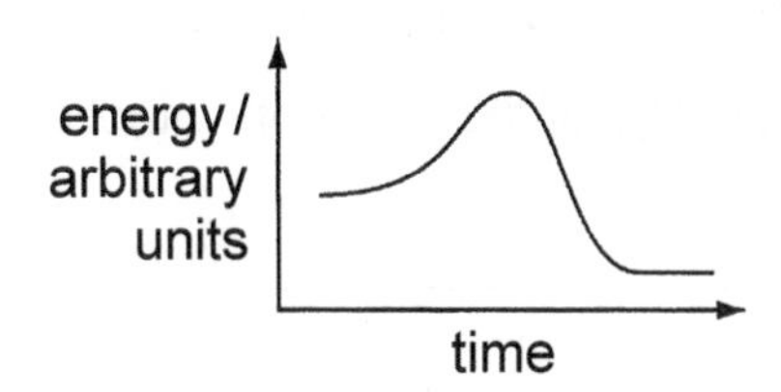

Which shape is the graph if the enzyme specific to this reaction is added at the start of the reaction?

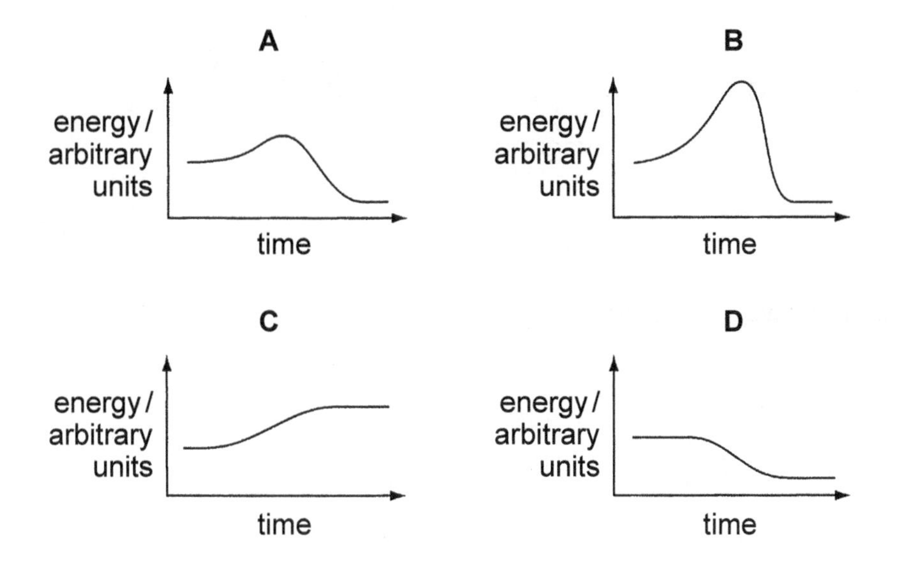

( )
[N11/1/6]

7. Three statements about water are listed.

1 Water cools a surface from which it evaporates.
2 Water is used as a solvent for many chemicals.
3 Water is involved in many metabolic reactions.

Which statement(s) make water suitable to use in a blood transport system?

A 1 and 2
B 1 and 3
C 2 only
D 3 only

( )
[N10/1/4]

8. The graph shows changing energy levels during a reaction, with and without the presence of the enzyme specific to this reaction.

What is the activation energy of the reaction without the enzyme?

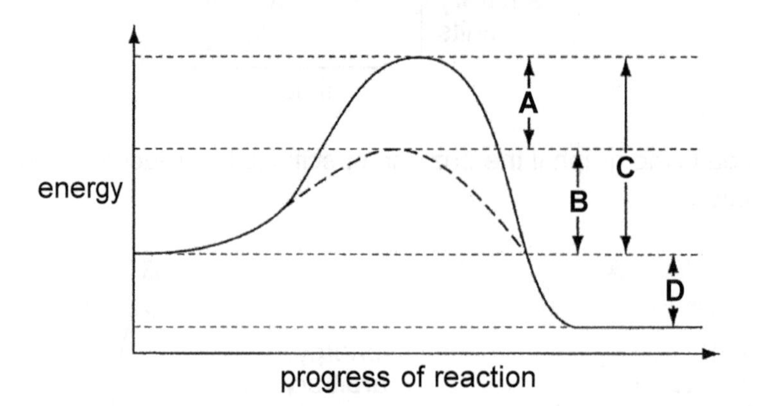

( )

[N10/1/5]

9. The diagram shows a cross-section of a leaf.

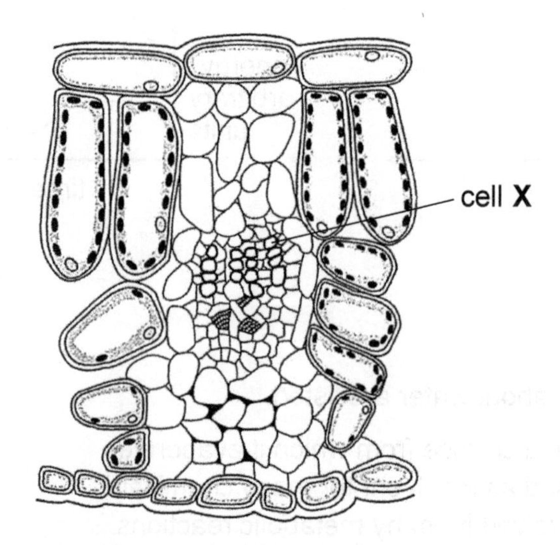

Samples of the contents of cell **X** are tested.
Which results will be obtained?

	Benedict's solution	iodine in a potassium iodide solution
A	+	+
B	+	–
C	–	+
D	–	–

key
+ = positive result
– = negative result

( )

[N10/1/6]

10. A food gives the following results on testing:

a purple colour in the biuret test;
a blue colour when heated with Benedict's solution;
a yellow colour with iodine;
a white emulsion with ethanol.

Which nutrients does the food contain?

A fat and protein
B protein and reducing sugar
C reducing sugar and starch
D starch and fat

( )
[N09/1/4]

11. The diagram shows an experiment to investigate the action of amylase on a 1.0 g cube of potato. The potato contains starch.

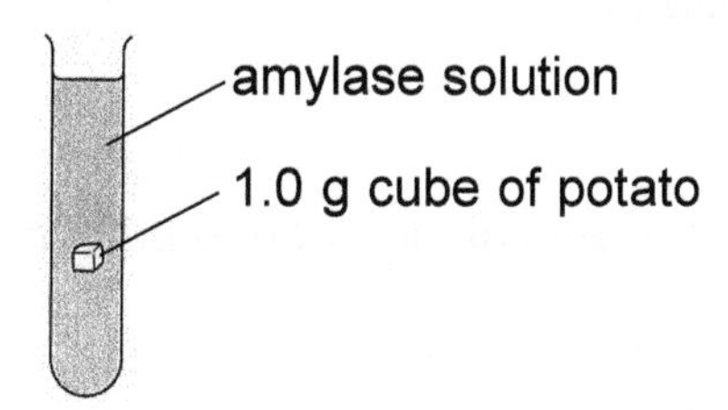

After 15 minutes at 20°C, 0.25 g of starch has been converted to sugar.
The experiment is repeated at 30°C.
How much starch is converted to sugar?

A 0.15 g
B 0.25 g
C 0.50 g
D 1.0 g

( )
[N09/1/5]

12. The graph shows the results of an experiment on an enzyme-controlled reaction.

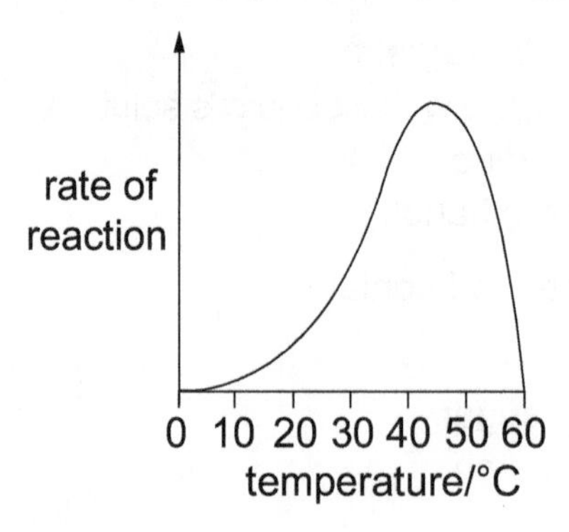

Which statement explains the shape of the graph above 40°C?

A The activation energy is exceeded.
B The enzyme-substrate complex does not separate.
C The shape of the active site of the enzyme is altered.
D The substrate is denatured. ( )

[N09/1/6]

13. Which property of water is essential for blood to be pumped by the human heart?

A It has a high specific heat capacity.
B It is an excellent solvent.
C It is most dense at +4°C.
D It is incompressible. ( )

[N08/1/4]

14. The diagram represents a large molecule, synthesised by joining smaller molecules of two types together.

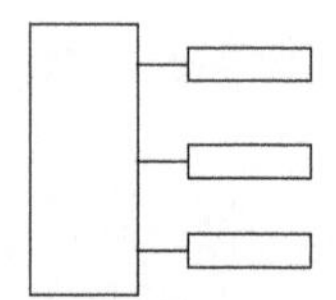

Which line names the large molecule and its components?

	molecule	amino acid	fatty acid	glucose	glycerol
A	glycogen	✓	×	✓	×
B	lipid	×	✓	×	✓
C	polypeptide	✓	×	×	✓
D	starch	×	✓	✓	×

key
✓ = component present
× = component absent

( )

[N08/1/5]

15. Four test-tubes were set up as shown.

5 drops of dilute hydrochloric acid are added to tubes **1**, **2** and **3**.

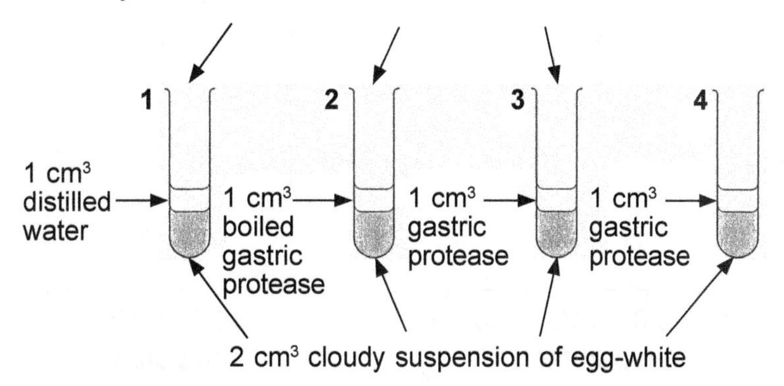

The contents of all four tubes were stirred and the tubes were then placed in a water-bath at 37°C for 20 minutes.
What is the result?

	tube number			
	1	**2**	**3**	**4**
A	clear	clear	clear	clear
B	clear	cloudy	cloudy	clear
C	cloudy	cloudy	clear	cloudy
D	cloudy	cloudy	cloudy	clear

( )
[N08/1/6]

16. The diagrams represent an enzyme molecule and its substrate.

Which diagram shows these molecules after they are heated to 100°C?

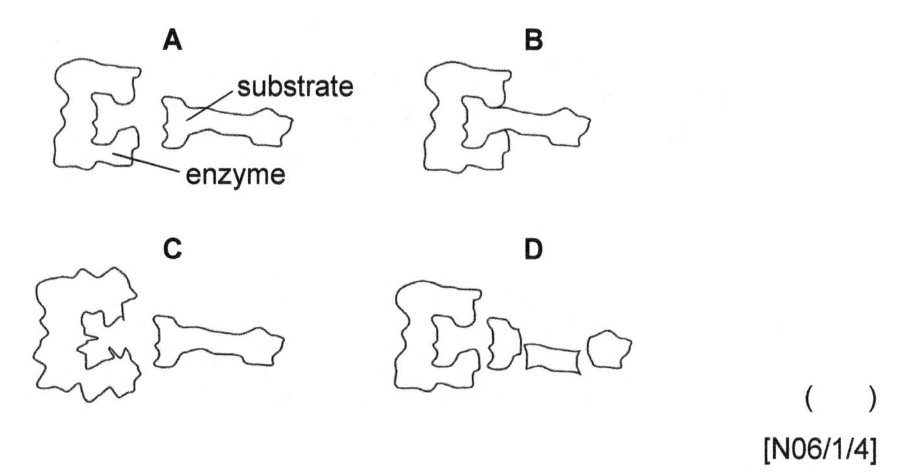

( )
[N06/1/4]

17. Which graph shows how an enzyme catalysed reaction in the alimentary canal varies with temperature?

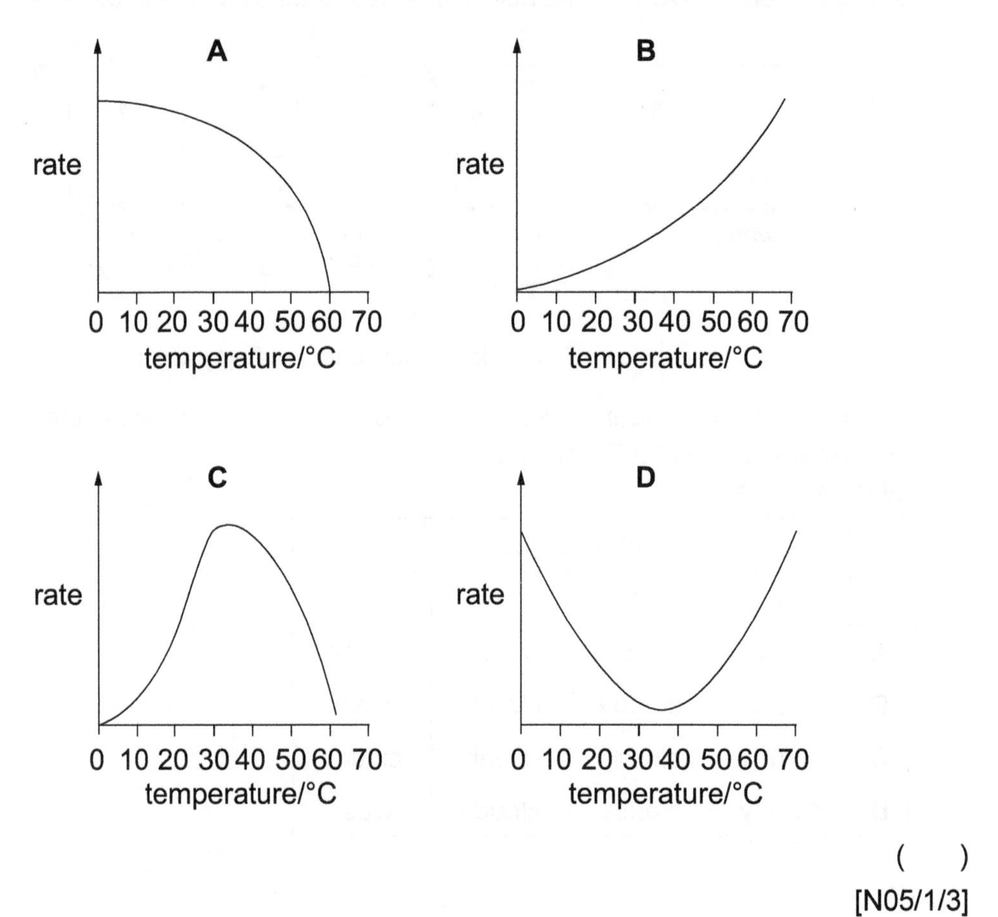

( )

[N05/1/3]

18. Four different foods were tested as shown below and the test results were recorded as positive (+) or negative (–).

Which food contained both glucose and oil?

food	Benedict's test	biuret test	ethanol emulsion test	iodine test
A	+	+	–	–
B	+	–	+	–
C	–	+	–	+
D	–	–	+	+

( )

[N05/1/7]

19. The diagram represents stages in the breakdown of starch to maltose by the enzyme amylase.

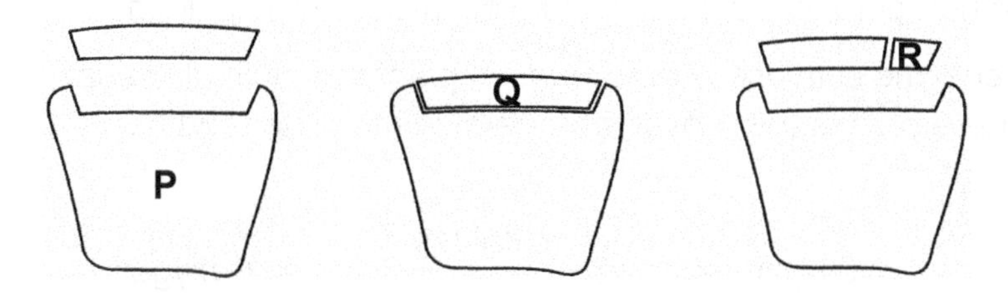

Which are the correct labels?

	starch	maltose	amylase
A	**P**	**Q**	**R**
B	**Q**	**R**	**P**
C	**R**	**P**	**Q**
D	**R**	**Q**	**P**

( )

[N05/1/10]

STRUCTURED QUESTIONS . . . .

1. **(a)** Hydrogen peroxide is a toxic chemical produced in plant and animal cells.

The enzyme catalase is also found in plant and animal tissues.
Catalase breaks down hydrogen peroxide to water and oxygen.

$$\text{hydrogen peroxide} \xrightarrow{\text{catalase}} \text{water + oxygen}$$

The mixture of water and oxygen forms a froth.

An investigation was carried out into the factors affecting the action of catalase.
The figure shows the results of the investigation.

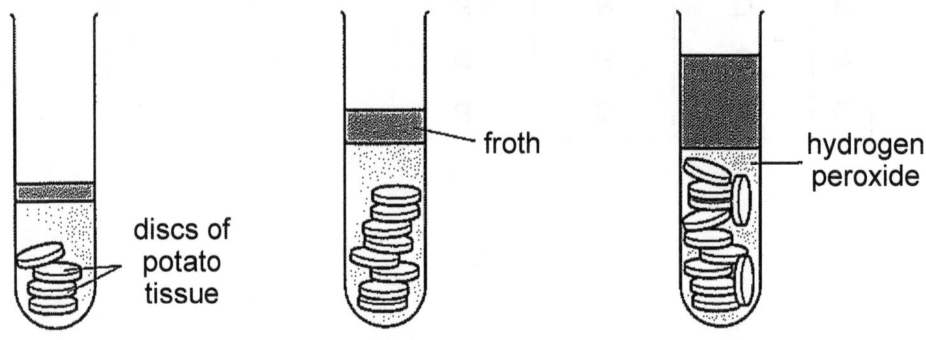

(i) State the factor being investigated.

....[1]

(ii) State two factors that need to be kept constant in this investigation.

1. ....

2. ....[2]

(iii) Suggest a suitable control for this investigation.

....

....[1]

(b) State a conclusion that can be made from the results shown in the figure.

....

.... [1]

[N12/2/6]

2. **(a)** **(i)** The figure represents part of a protein molecule.

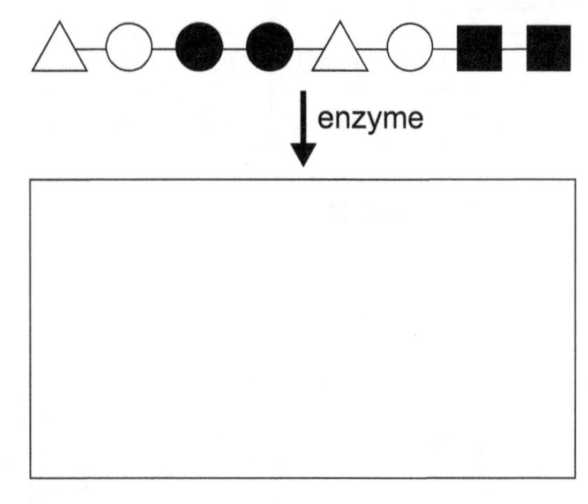

Draw a diagram in the box to show all the molecules present after the complete digestion of this part of the protein. [2]

(ii) Name the type of enzyme which digests the protein.

....

....[1]

(b) Describe how the activity of this enzyme is affected by temperature.

....

....

....

....

.... [4]

[N08/2/3]

3. The table shows the results of an investigation to find out how temperature affects the activity of a protease enzyme.

temperature / °C	0	10	20	30	40	50	60
mg of product formed	1.8	6.2	14.0	31.6	40.0	17.4	0.8

(a) Plot this data on the graph paper. [3]

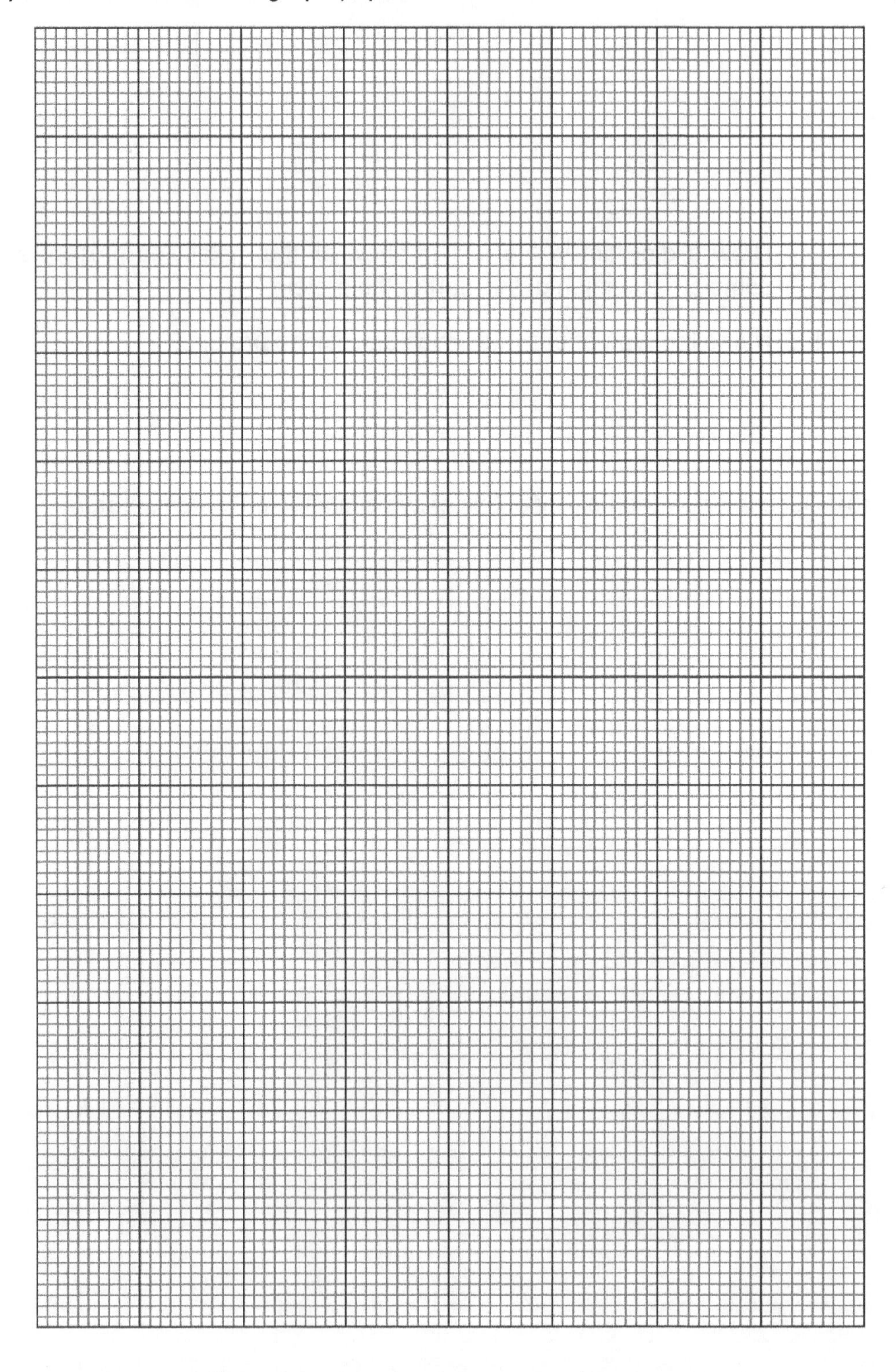

(b) Describe the relationship between enzyme activity and temperature.

....

....

....

....

.... [4]

(c) Explain the results above 40°C.

....

....

....

.... [3]

[N07/2/2]

FREE RESPONSE QUESTIONS. . . . .

1. **(a)** Define the term *enzyme*. [3]

   (b) Sketch a graph to show the effect of pH on the rate of an enzyme-catalysed reaction. Label the axes. [3]

[N11/2/10 Either (a), (b)]

TOPIC 4

Nutrition in Humans

MULTIPLE CHOICE QUESTIONS . . . .

1. Which section of the diagram represents the functions carried out by the liver?

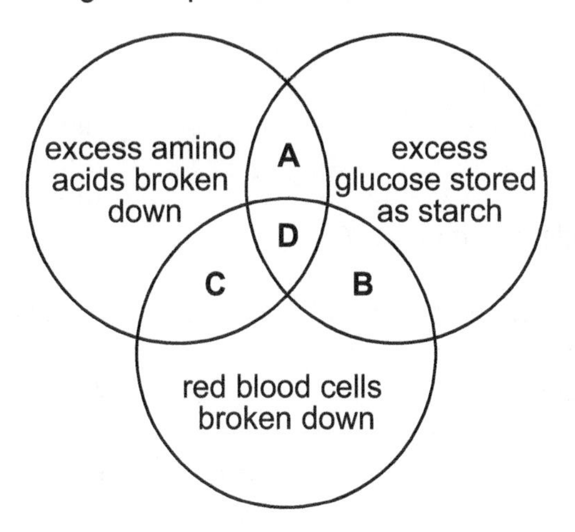

( )

[N12/1/7]

2. The table gives the results of an investigation into the rates of absorption of digested food in the ileum.

The experiment was carried out in the presence of oxygen and then repeated without oxygen, with the same amount of food being tested each time.

product of digestion	rate of absorption in the presence of oxygen / arbitrary units	rate of absorption in the absence of oxygen / arbitrary units
amino acids	5.4	1.2
fatty acids	1.8	1.7
glucose	6.8	4.2
glycerol	4.2	4.1

Which conclusion can be drawn from these results?

A Amino acids are absorbed by active transport only.
B Glycerol is absorbed mainly by diffusion.
C Glycerol is absorbed more slowly than fatty acids.
D More glucose is absorbed by active transport than by diffusion. ( )

[N12/1/8]

3. In which two body organs does the greatest amount of reabsorption of water take place?

A colon and kidneys
B duodenum and colon
C kidneys and liver
D liver and duodenum

( )

[N11/1/7]

4. The diagram shows a food bolus moving down the oesophagus.

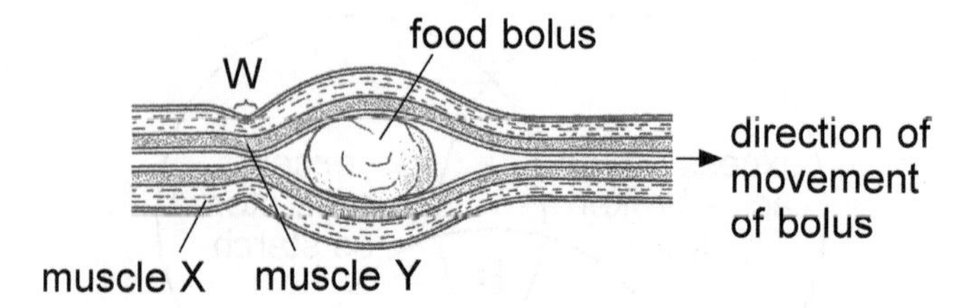

Which row identifies the muscles and their actions at region W?

	muscle X		muscle Y	
	muscle type	muscle action	muscle type	muscle action
A	circular	contracting	longitudinal	relaxing
B	circular	relaxing	longitudinal	contracting
C	longitudinal	contracting	circular	relaxing
D	longitudinal	relaxing	circular	contracting

( )

[N11/1/8]

5. Which function of the liver is correctly paired with the chemical involved?

	function	chemical
A	deamination	glycogen
B	detoxification	alcohol
C	excretion	urea
D	storage	amino acids

( )

[N10/1/7]

6. The diagram shows the human alimentary canal with labels for the functions of some of its parts.

Which label is correct?

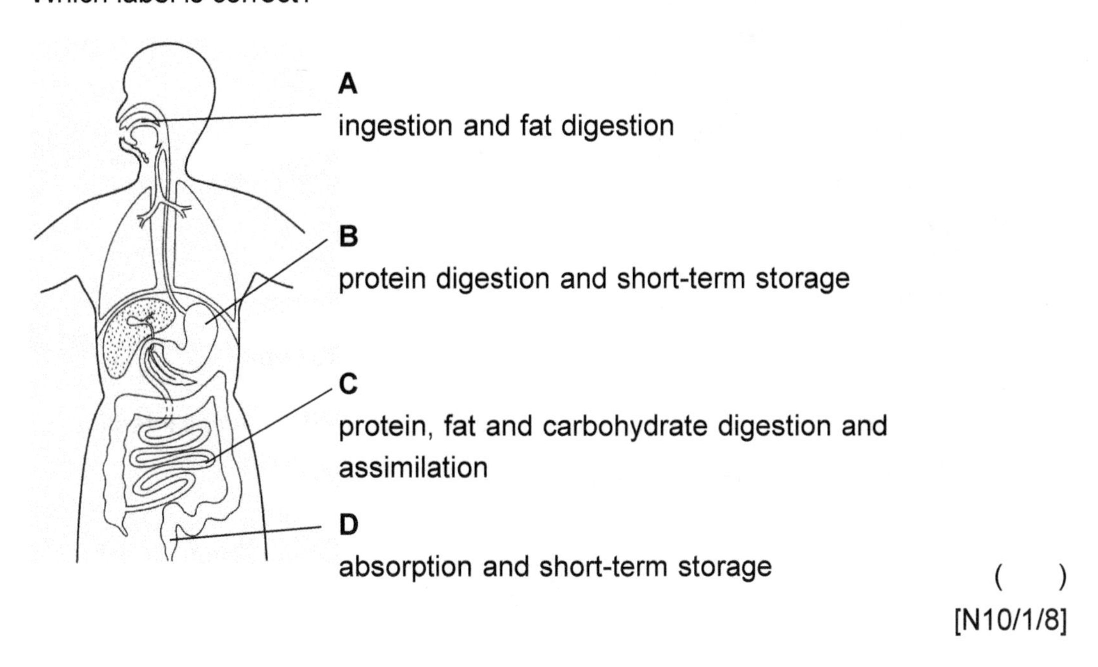

( )

[N10/1/8]

7. The diagram shows the human digestive system.

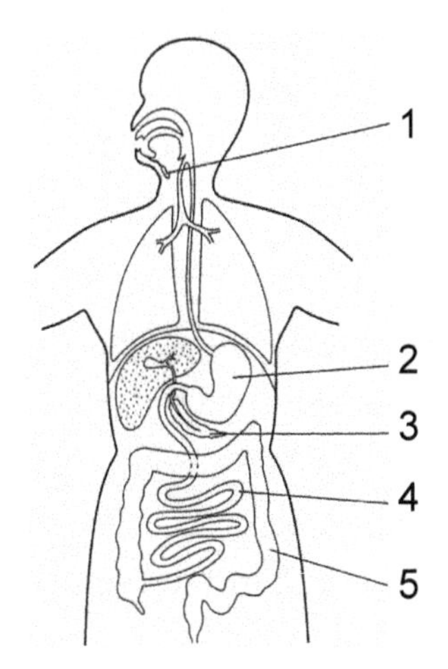

Which structures make the enzyme that breaks down starch?

A 1 and 3 only
B 1 and 5 only
C 2 and 3 only
D 2 and 4 only

( )

[N09/1/7]

8. The statements are about how the liver deals with excess glucose.

1 It is converted to amino acids.
2 It is converted to fats.
3 It is deaminated to produce urea.
4 It is stored as glycogen.

Which statements are correct?

A 1 and 2 only
B 1, 2 and 3
C 2 and 4
D 4 only

( )

[N09/1/8]

9. The diagram shows the internal structure of a villus from the small intestine.

Which structure receives absorbed lipids?

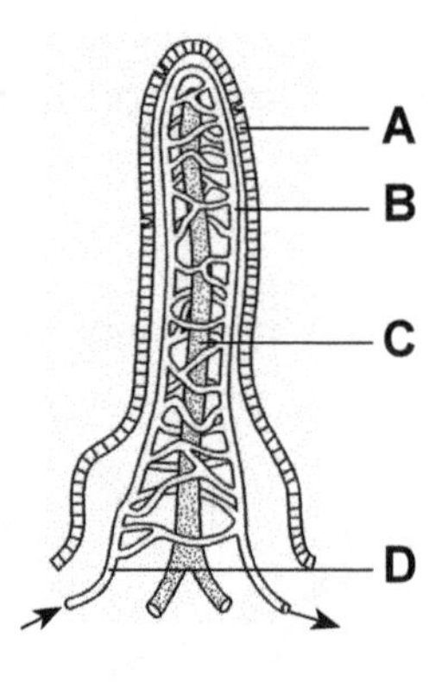

( )

[N08/1/7]

10. Two hours after a student ate a meal of protein, fat and carbohydrate, the contents of the blood in the hepatic portal vein and the hepatic vein were compared.

Which comparison is correct?

	blood in the hepatic portal vein	blood in the hepatic vein
A	less amino acids	more amino acids
B	less red blood cells	more red blood cells
C	more glucose	less glucose
D	more urea	less urea

( )

[N08/1/8]

11. Why does chewing food help digestion?

A The food is easier to swallow.
B The food has a better taste.
C The food has an increased surface area.
D The food is softer.

( )

[N07/1/7]

12. In the small intestine, what is the main function of the villi?

A to cause food to be swept along more swiftly
B to cause more efficient absorption of digested food
C to remove water from undigested food
D to secrete enzymes

( )

[N07/1/8]

13. The diagram shows part of the human alimentary canal.

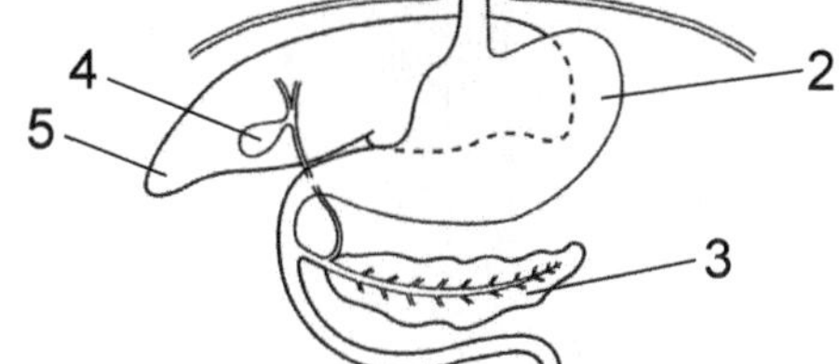

Which two structures produce substances involved in the digestion of fat?

A 1 and 4
B 2 and 3
C 3 and 5
D 4 and 5

( )

[N07/1/9]

14. Which line in the table states a function of the blood in the hepatic portal vein?

	chemical carried	from	to
A	carbon dioxide	liver	vena cava
B	amino acids	ileum	liver
C	glycerol	ileum	liver
D	urea	liver	kidney

( )

[N07/1/10]

15. The diagram shows part of the human digestive system.

Which part secretes an acidic digestive juice containing a protease?

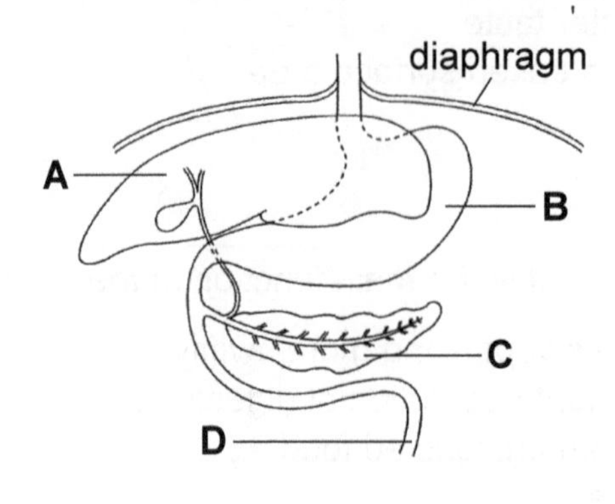

( )

[N06/1/10]

16. The diagram shows a section through a villus.

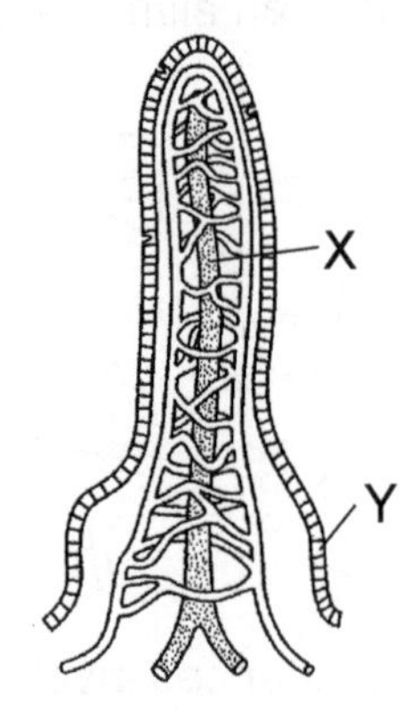

What is a function of structure X and structure Y?

	X	Y
A	to absorb amino acids	to digest starch
B	to carry blood	to secrete mucus
C	to transport fats	to secrete enzymes
D	to transport glucose	to help peristalsis

( )

[N06/1/11]

STRUCTURED QUESTIONS . . . .

1. The figure shows an apparatus set up to investigate digestion.

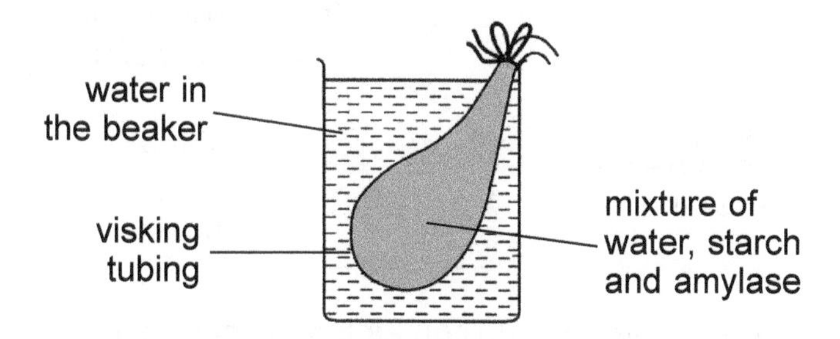

The apparatus was left at 30°C for four hours.
After this time the water in the beaker was tested and found to contain reducing sugars.

(a) (i) Explain why reducing sugars were found in the water in the beaker.

....

....

....

....[3]

(ii) The water in the beaker was also tested for starch.

State the result of this test and explain your answer.

....

....

....

....[2]

(b) State the property of the visking tubing which made these results possible.

....

.... [1]

[N11/2/1]

2. The figure shows the apparatus used in an investigation into the action of digestive enzymes.

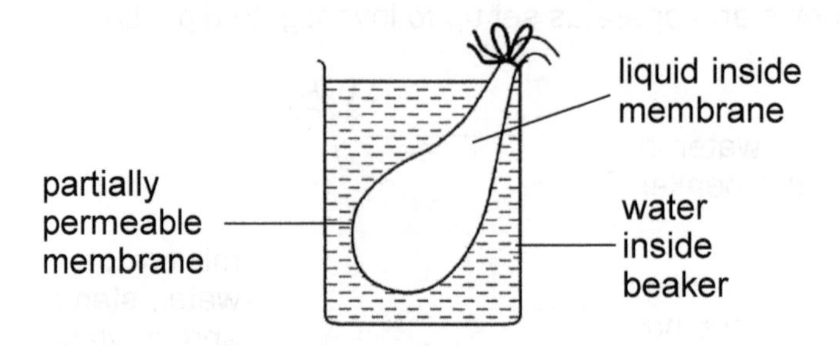

Five sets of apparatus were set up containing the substances shown in **Table (a)**. Each set of apparatus was left for 1 hour at 30°C.

Table (a)

apparatus set number	contents of liquid inside membrane
1	starch
2	starch and amylase
3	starch and boiled amylase
4	protein and amylase
5	protein and amino acids

The expected results are shown in the **Table (b)**.

Table (b)

expected results	apparatus set number(s)
starch in tubing, no starch or sugar in water	
no starch in tubing, sugar in water	
protein in tubing, no amino acids in water	
protein in tubing, amino acids in water	

(a) Using the numbers 1 to 5, match the set(s) of apparatus to the expected results in **Table (b)**. [5]

(b) Describe the role of bile in the digestion of fats.

....

....

.... [2]

[N10/2/3]

3. In an investigation, a cloudy, white gel containing milk protein was prepared.

The gel was poured into Petri dishes and allowed to set.

Cavities were made in the gel and various liquids were poured into the cavities as shown in **Fig. (a)**.

- 1 contained pepsin and hydrochloric acid
- 2 contained pepsin
- 3 contained boiled pepsin
- 4 contained water

Fig. (b) shows the results after 24 hours.

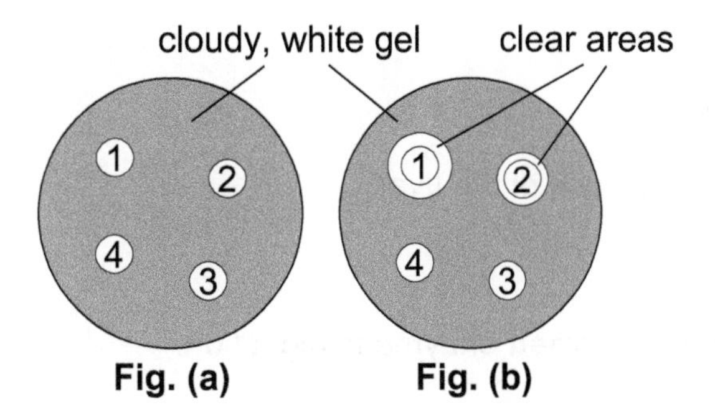

Fig. (a) **Fig. (b)**

(a) Suggest what caused the clear areas around cavities 1 and 2.

....

....

....

.... [3]

(b) Explain the difference in the results between cavities 1 and 2.

....

.... [1]

(c) Explain why no clear area developed around cavity 3.

....

.... [1]

(d) State the purpose of cavity 4.

....

.... [1]

[N09/2/6]

FREE RESPONSE QUESTIONS. . . . .

1. **(a)** Describe the digestion of protein in the body. [3]

(b) Villi are found in the digestive system.

Describe the structure and function of a villus. [3]

(c) Describe the role of the liver in the metabolism of carbohydrate. [4]

[N12/2/10 Or]

2. **(a)** State the meaning of the terms:

(i) *digestion* [3]

(ii) *absorption.* [3]

(b) Describe the functions of the liver. [4]

[N11/2/10 Or]

3. Describe the role of a **named** enzyme in digestion. [4]

[N11/2/10 Either (c)]

4. Describe the possible harmful effects on the body of alcohol consumption.

[4]

[N07/2/6(b)]

TOPIC 5

Nutrition in Plants

MULTIPLE CHOICE QUESTIONS . . . .

1. Which equation summarises photosynthesis?

A $6CO_2 + C_6H_{12}O_6 \rightarrow 6H_2O + 6O_2$
B $6CO_2 + 6H_2O \rightarrow C_6H_6O_6 + 6O_2$
C $6H_2O + 6CO_2 \rightarrow C_6H_{12}O_6 + 6O_2$
D $6O_2 + C_6H_{12}O_6 \rightarrow 6H_2O + 6CO_2$ ( )

[N12/1/9]

2. The graph shows the rate of photosynthesis plotted against an unknown factor.

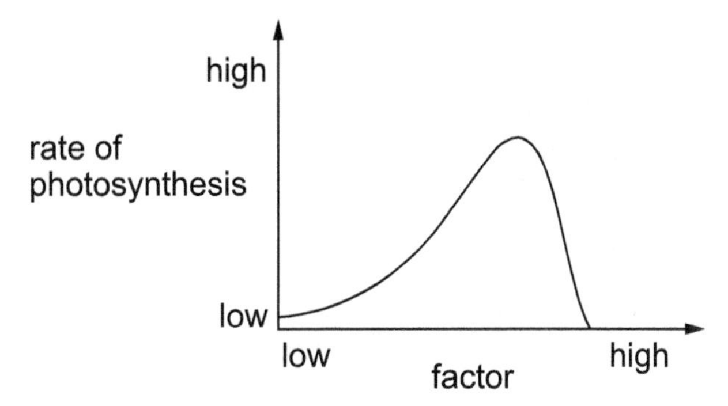

Which factor is limiting the rate of photosynthesis as shown in the graph?

A carbon dioxide concentration
B light intensity
C number of chloroplasts
D temperature ( )

[N12/1/10]

3. The diagram shows a section through a leaf.

In which layer will there be **no** conversion of light energy to chemical energy?

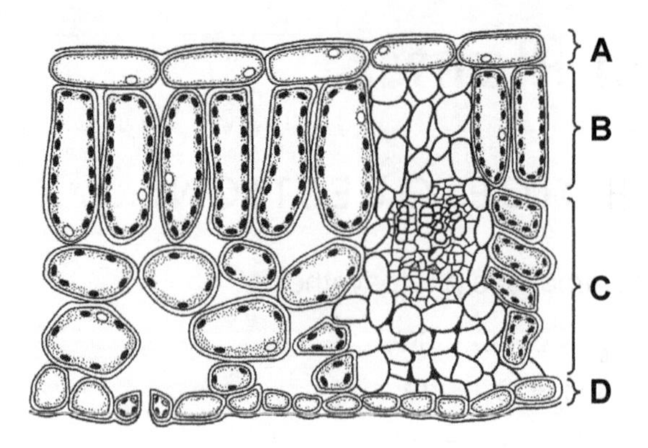

( )

[N11/1/9]

4. Some students investigated gaseous exchange in a green plant. The rate of oxygen production was plotted against carbon dioxide concentration.

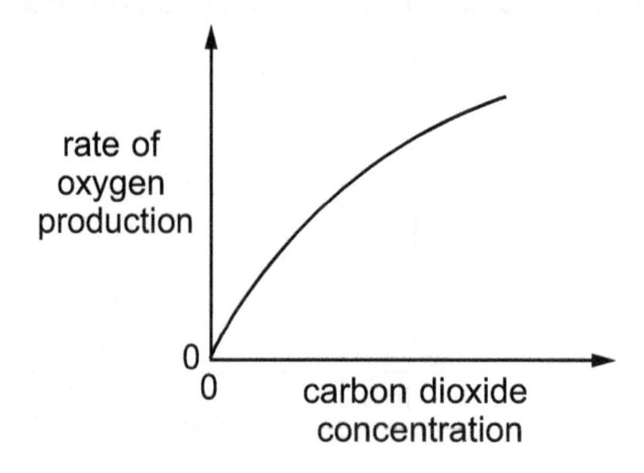

What explains these results?

A Carbon dioxide controls the rate of respiration.

B Carbon dioxide controls the rate of photosynthesis.

C Oxygen controls the rate of photosynthesis.

D Oxygen controls the rate of respiration.

( )

[N11/1/10]

5. Which chemical change takes place in green plants but **not** in animals?

A glucose → cellulose
B glucose → glycogen
C glycogen → glucose
D glycogen → cellulose

( )
[N10/1/9]

The photomicrograph shows a cross-section through a leaf. Use the photomicrograph to answer question 6.

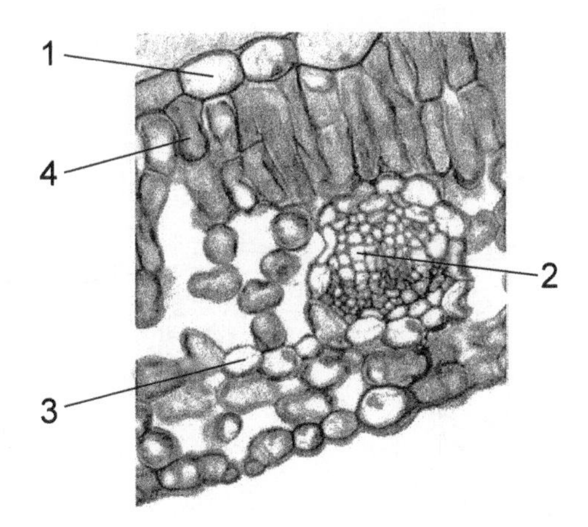

6. Which labelled cell has the function shown?

	cell	function
A	1	absorption of light
B	2	transport of sucrose
C	3	transport of water
D	4	synthesis of carbohydrate

( )
[N10/1/10]

7. What uses stomata as its route into the leaf?

A carbon dioxide
B ions
C sunlight
D water

( )
[N10/1/11]

8. Plants use carbon dioxide and water to carry out photosynthesis.

Which description is correct?

A Carbon dioxide and water enter the roots and move through the xylem to each leaf.

B Carbon dioxide and water enter a leaf through stomata and diffuse into leaf cells.

C Carbon dioxide enters a leaf through stomata and water enters a leaf from the xylem.

D Carbon dioxide enters a leaf through stomata and water enters a leaf from the phloem.

( )

[N09/1/9]

9. The diagram shows part of a cross-section of a leaf.

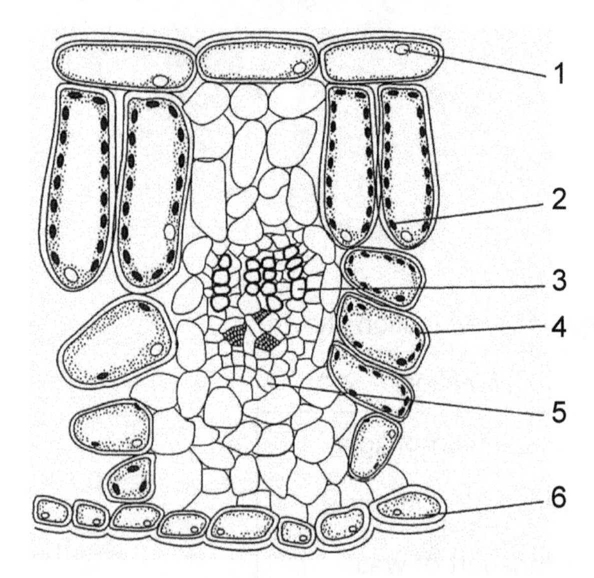

Which labelled structures convert light energy to chemical energy?

A 1 and 2

B 1 and 6

C 2 and 4

D 3 and 5

( )

[N09/1/10]

10. The diagram shows a section through a leaf of a green plant.

During exposure to bright light, which region contains most starch?

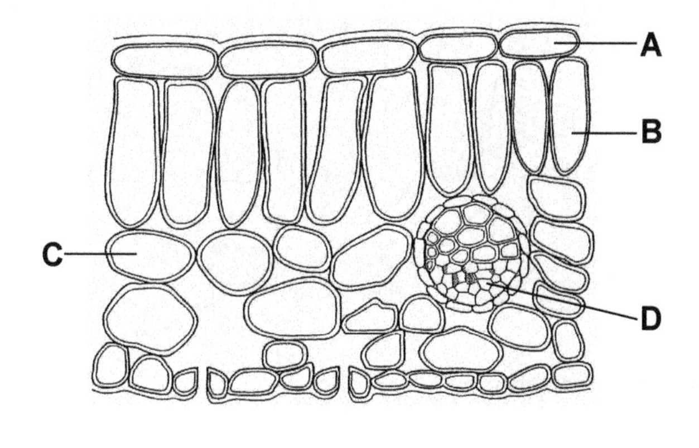

( )

[N08/1/9]

11. The graph shows the rate of photosynthesis plotted against an unknown factor.

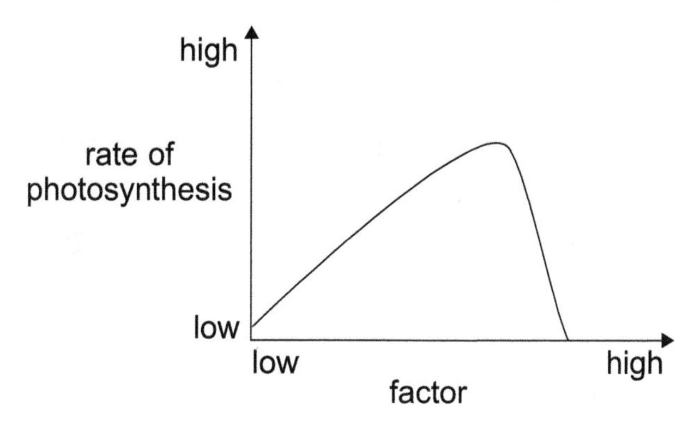

Which factor is limiting the rate of photosynthesis as shown in the graph?

A carbon dioxide concentration
B light intensity
C number of chloroplasts
D temperature

( )

[N08/1/10]

12. The diagram shows part of a transverse section of a leaf.

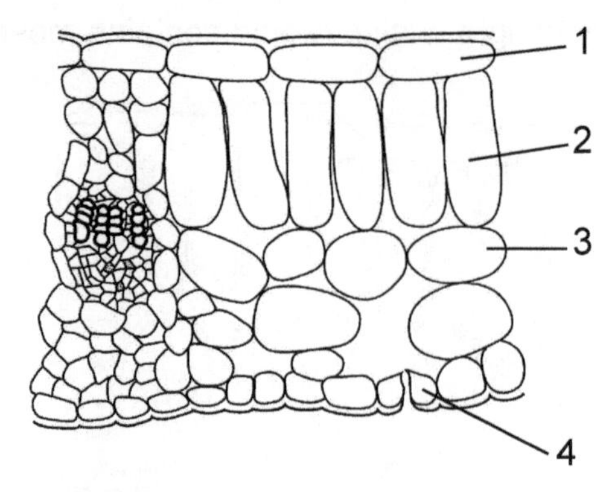

Which cells contain chloroplasts?

A 1 and 2 only
B 2 only
C 2 and 3 only
D 2, 3 and 4 only

( )
[N07/1/4]

13. Which graph shows the effect of temperature on the release of oxygen by an illuminated water plant, such as *Elodea*?

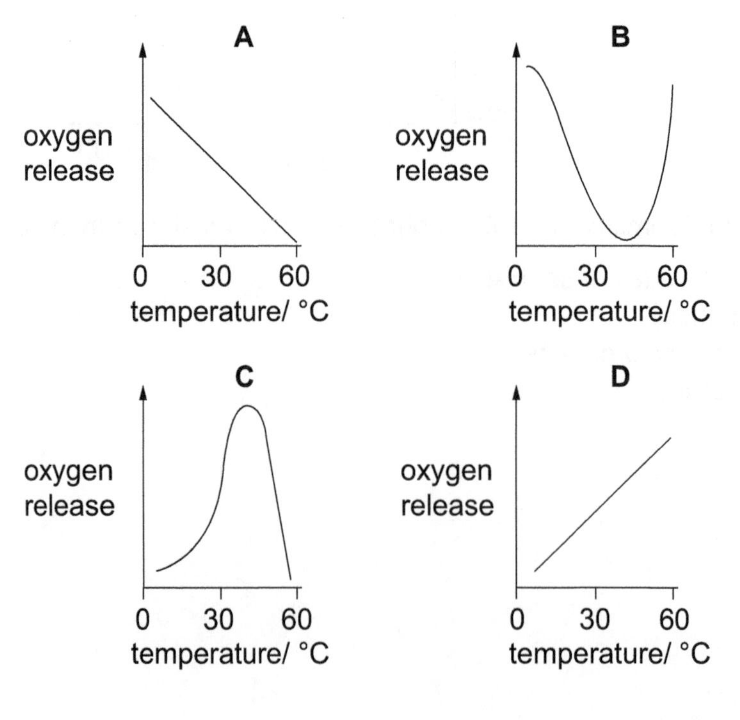

( )
[N07/1/5]

14. The graph shows the effect of different wavelengths of light on processes taking place in green plants.

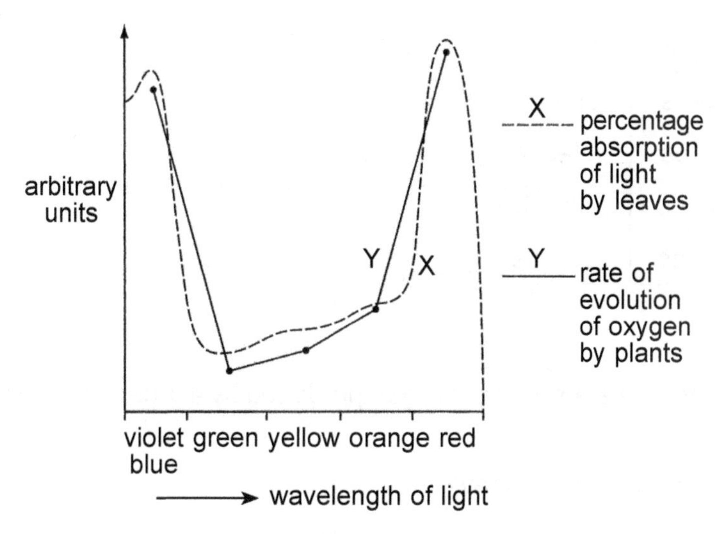

What can be deduced from the solid line Y?

A Photosynthesis is least active in green light and most active in blue and red light.
B Photosynthesis is most active in green light and least active in blue and red light.
C Respiration is least active in green light and most active in blue and red light.
D Respiration is most active in green light and least active in blue and red light.

( )

[N07/1/6]

15. The graph shows the effect of changing light intensity on the rate of photosynthesis in a plant, at two different concentrations of carbon dioxide.

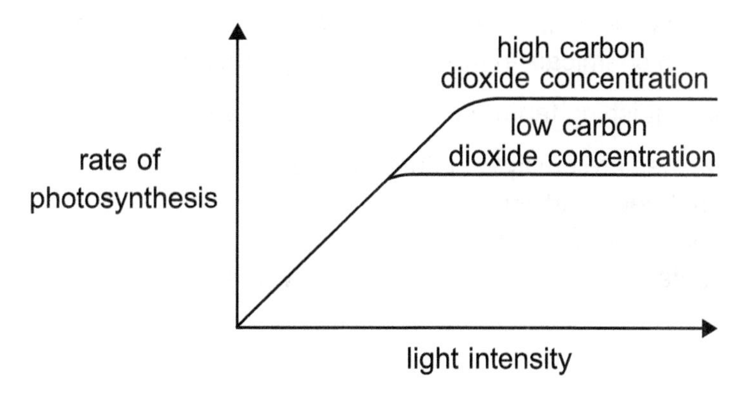

What conclusions can be drawn from the graph?

A At high light intensities, carbon dioxide limits the rate of photosynthesis.
B At low light intensities, light intensity has no effect on photosynthesis.
C Light intensity limits the rate of photosynthesis.
D When carbon dioxide concentration is low, the plant cannot photosynthesise.

( )

[N06/1/5]

16. When is carbon dioxide absorbed and when is it released by an ecosystem, such as a tropical rainforest?

	daylight	darkness
A	absorbs	absorbs
B	absorbs	releases
C	releases	absorbs
D	releases	releases

( )
[N06/1/6]

17. The graph shows the amount of oxygen produced by a green plant, growing outdoors, during a 24-hour period.

Which letter represents midday?

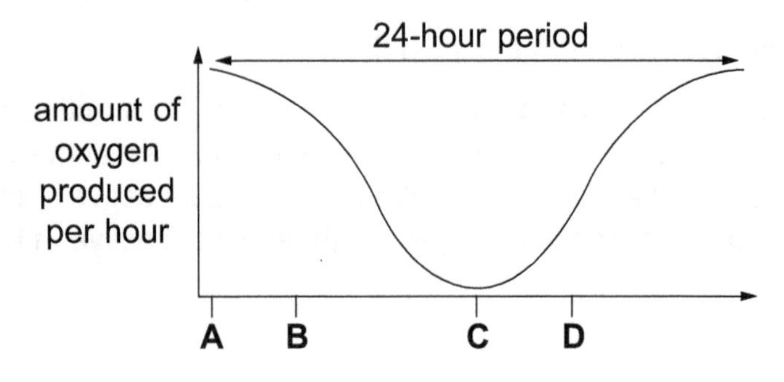

( )
[N06/1/7]

18. Some organisms live in the dark at the bottom of the seas and, to synthesise glucose, use energy from chemicals in the very hot water that comes out of volcanoes.

What is a distinguishing feature of these organisms?

A Their enzymes are easily denatured by heat.
B They do not need carbon dioxide.
C They do not need to be green.
D They all obtain energy only by being carnivores.

( )
[N06/1/8]

19. The diagram shows some apparatus used in investigating gas exchange.

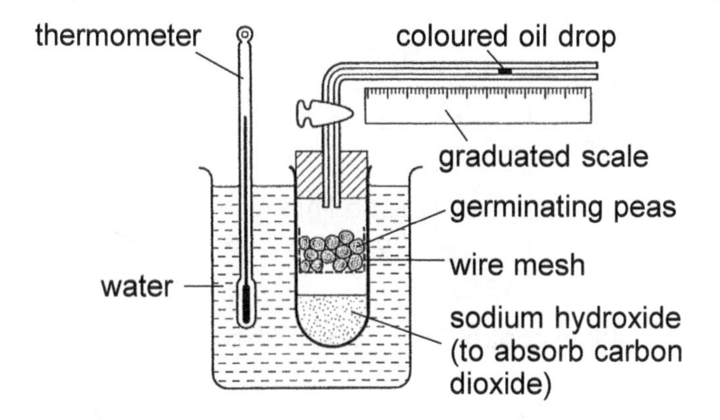

By measuring the movement of the oil drop in the apparatus, what can be investigated?

A carbon dioxide released during germination
B heat released during germination
C oxygen used during germination
D water produced during germination ( )

[N06/1/18]

20. Some students investigated the photosynthesis of a water plant. The rate of oxygen production was plotted against carbon dioxide concentration.

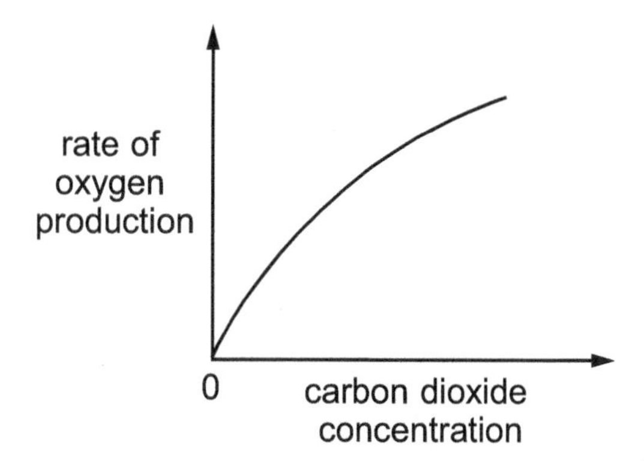

What explains these results?

A Carbon dioxide controls the rate of respiration.
B Carbon dioxide controls the rate of photosynthesis.
C Oxygen controls the rate of photosynthesis.
D Oxygen controls the rate of respiration. ( )

[N05/1/5]

21. The diagram shows an experiment to find out whether carbon dioxide is needed for photosynthesis.

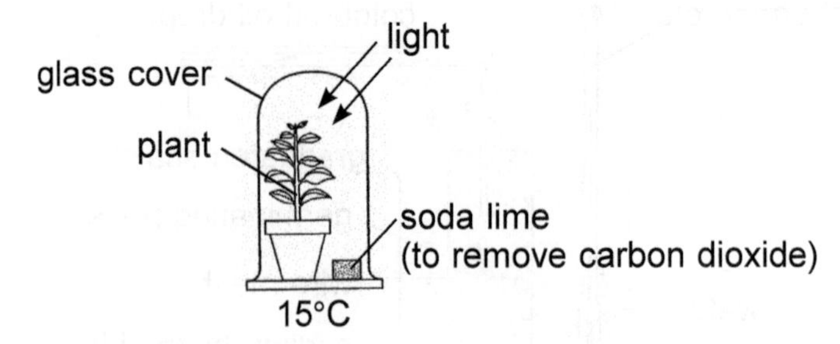

What is the most suitable control for this experiment?

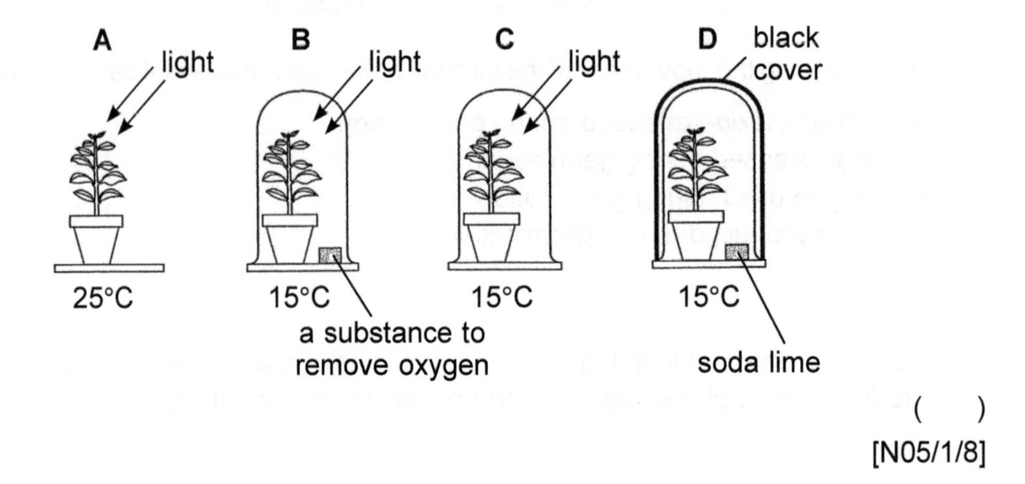

( )

[N05/1/8]

STRUCTURED QUESTIONS . . . .

1. **(a)** **(i)** In the space provided, write the word equation for photosynthesis.

(ii) State the name of the structures in which photosynthesis takes place.

....[1]

(iii) State the type of cell in which most of these structures are found.

....[1]

(b) **Fig. (a)** shows a scanning electron micrograph image of the lower surface of a leaf of *Hibiscus schizopetalus*.

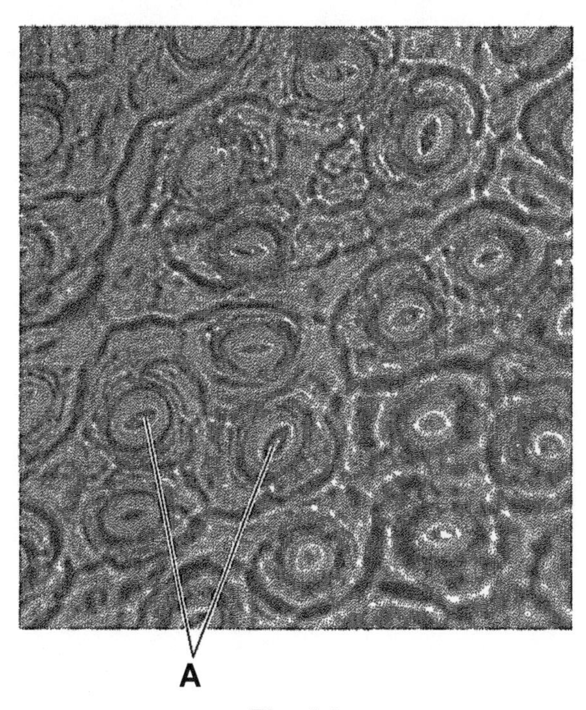

Fig. (a)

(i) Identify the structures labelled **A** in **Fig. (a)**.

....[1]

(ii) Describe the function of the structures labelled **A**.

....

....

....

....[2]

(c) **Fig. (b)** shows the effect of three factors on the rate of photosynthesis represented by lines **A**, **B** and **C**.

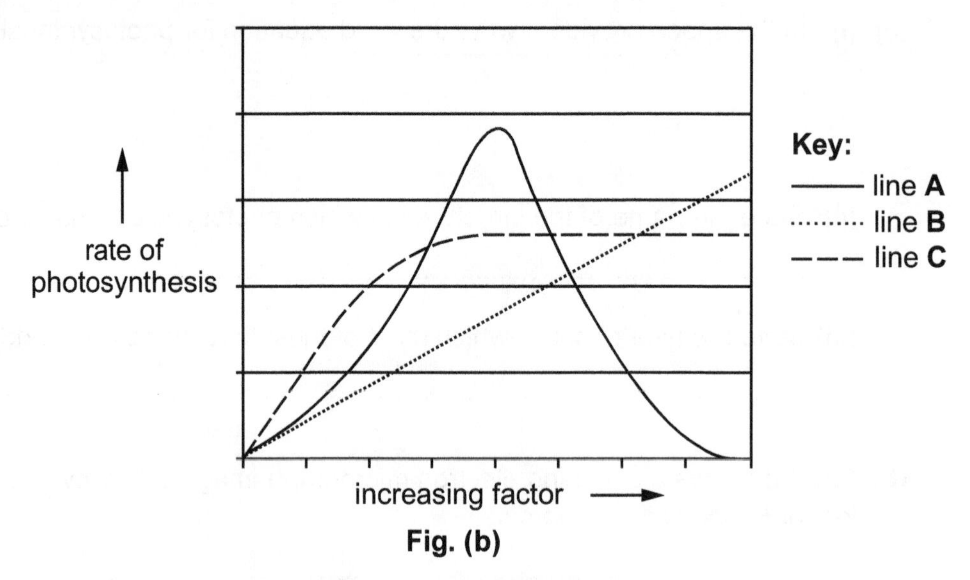

Fig. (b)

State which line, **A**, **B** or **C** represents the effect of increasing the concentration of carbon dioxide.

....[1]

[N12/2/4]

2. **Fig. (a)** shows a cell from the leaf of a plant.

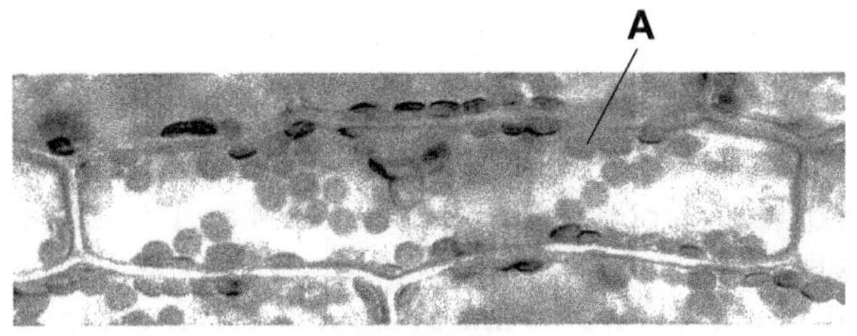

Fig. (a)

(a) **(i)** Name the structure labelled **A**.

....

....[1]

(ii) Name the gases which diffuse in and out of structure **A** during the day.

in ....

out ....

[2]

(b) **Fig. (b)** shows the effect of temperature on photosynthesis.

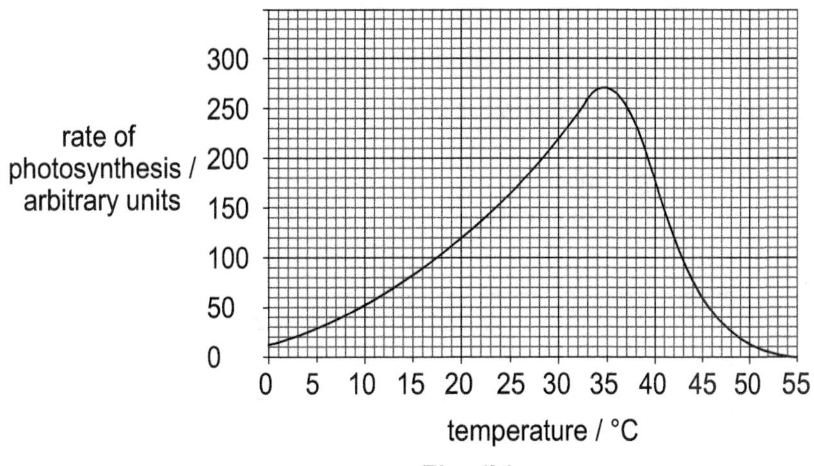

Fig. (b)

State and explain the relationship between temperature and photosynthesis.

....

....

....

....

....

.... [4]

[N11/2/6]

3. **(a)** State the word equation for photosynthesis.

....

.... [2]

(b) The rate of oxygen production in two plants, **X** and **Y**, was measured at different light intensities.
One of the plants grows in shady conditions and the other plant grows in sunny conditions.
The figure shows the results for plant **X**.

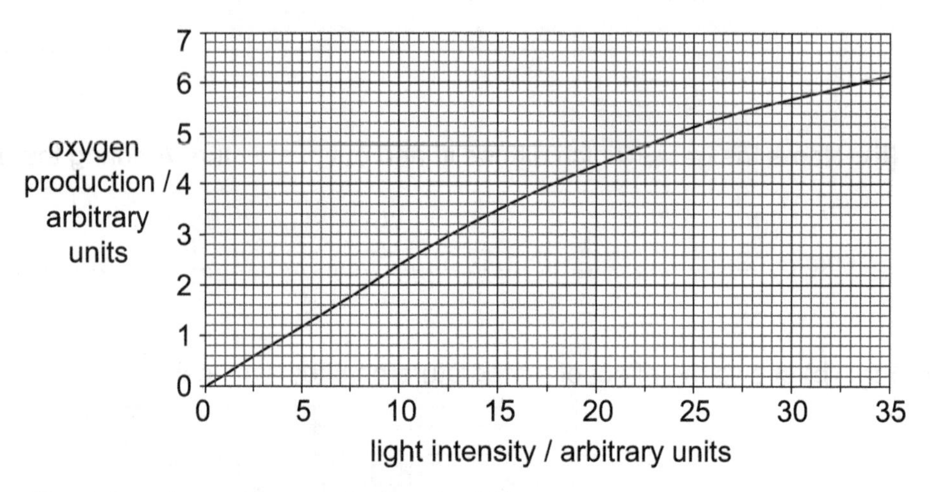

The table shows the results for plant **Y**.

light intensity / arbitrary units	oxygen production / arbitrary units
0	0.0
5	2.1
10	2.9
15	3.3
20	3.5
25	3.6
30	3.6
35	3.6

(i) Plot the data for plant **Y** on the figure. [2]

(ii) Describe the differences in the oxygen production between plants **X** and **Y**.

....

....

....[3]

(iii) Which of the two plants grows in the shade?
Explain your answer.

....

....

....[2]

[N08/2/6]

4. The figure shows a section through *Hydra*, a simple animal that lives in water. Part of its body wall has been enlarged to show cell detail.

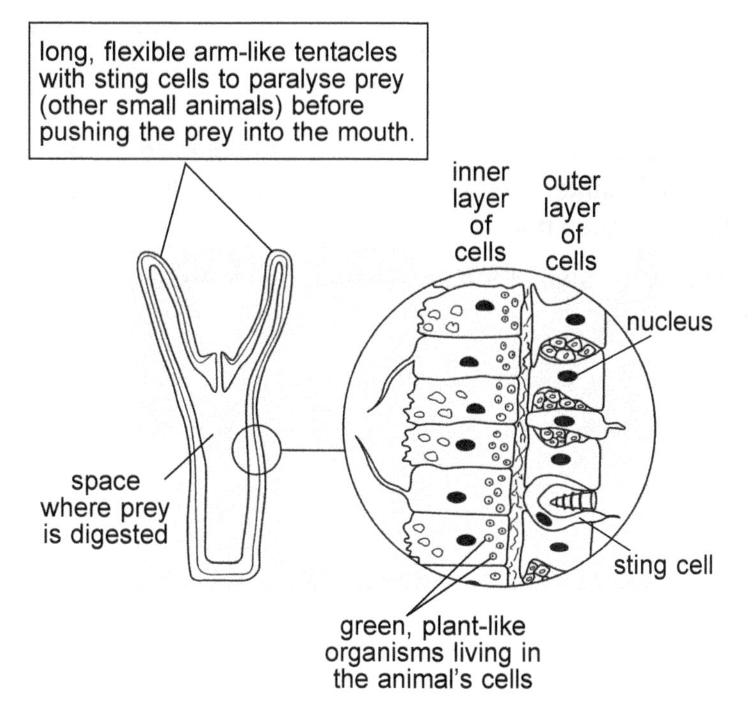

(a) In some of the cells are small plant-like organisms which are green because they contain the same pigment as green plants.

(i) Name the pigment.

....[1]

(ii) Deficiency of which mineral ion would affect the growth of such green organisms by reducing the production of this pigment.

....

....[1]

(b) (i) The animal supplies the plant-like organisms with a gas essential for the plant-like organisms' food production.

Name the gas and the process for which it is used.

gas ....

process .... [2]

(ii) Suggest how the animal might benefit from the presence of these plant-like organisms.

....

....

....[2]

(c) There is another species of *Hydra* closely-related to the animal in the figure, which is much darker green in colour and has no sting cells. Unlike the animals in the figure, it does not eat other organisms.

Suggest how the animal is able to obtain the necessary nutrients for energy and for growth.

....

....

....

.... [3]

[N05/2/3]

FREE RESPONSE QUESTIONS. . . . .

1. **(a)** Describe the role of chlorophyll in photosynthesis. [2]

(b) The rate of photosynthesis in six tropical crop plants was measured when the plants were growing outside under normal conditions (rate **X**).
The measurements were repeated again when the plants were grown under controlled optimum conditions in a glasshouse (rate **Y**). The results are shown in the table.

crop plant	rate of photosynthesis (X) / μmol per m^2 per second	rate of photosynthesis (Y) / μmol per m^2 per second	difference in the rate of photosynthesis (Y – X) / μmol per m^2 per second
cassava	13.7	23.1	9.4
eucalyptus	18.4	26.0	7.6
maize	23.4	26.0	2.6
soya bean	18.3	25.1	6.8
sugar cane	24.0	26.8	2.8
sunflower	24.3	31.7	7.4

(i) Draw a bar chart of the difference (**Y – X**) in the rate of photosynthesis of each plant.

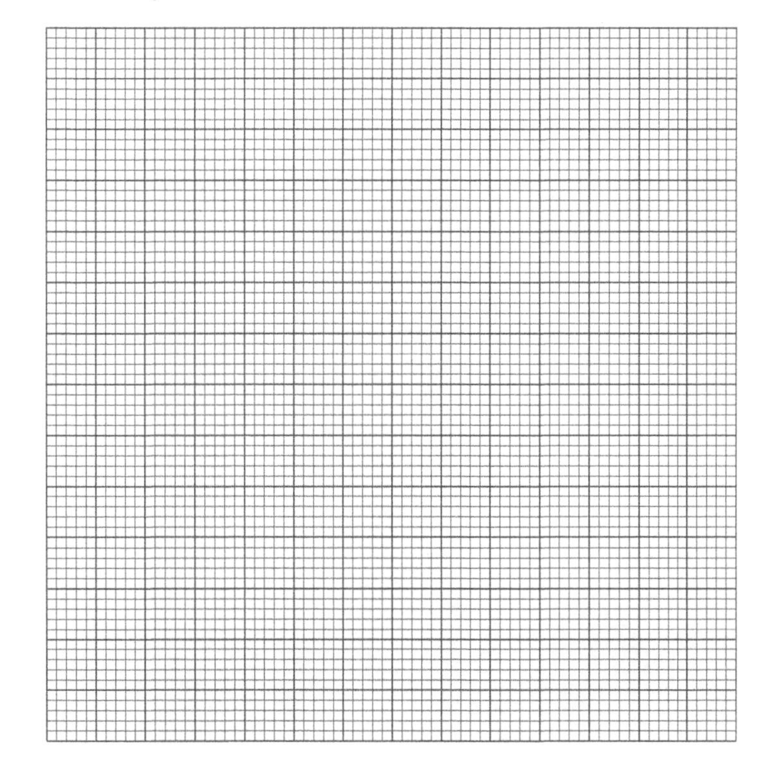

[4]

(ii) Which **two** crop plants show the greatest difference between rate **X** and rate **Y**? [2]

(iii) The measurements of the rate of photosynthesis (**X**) of the plants when grown outside are means of 10 readings.

Suggest a reason why mean measurements were used. [3]

(iv) Suggest two factors that were changed when the plants were grown in controlled optimum conditions. [2]

(c) The rate of photosynthesis can be measured by:

- calculating the rate per unit area of the leaf
  or
- calculating the rate per unit mass of the leaf.

Suggest why these measurements may give different results. [1]

[N10/2/8]

2. **(a)** Describe the process of photosynthesis. [4]

(b) Explain why animals are dependent on photosynthesis. [4]

(c) Describe the effects of increasing light intensity on photosynthesis. [2]

[N07/2/8 Either]

Transport in Flowering Plants

MULTIPLE CHOICE QUESTIONS . . . .

1. Where does **most** transpiration in a plant take place?

   A cuticle
   B mesophyll cells
   C stomata
   D xylem vessels ( )

   [N12/1/11]

2. Some insects use piercing mouthparts to obtain sugars from plants.

   The diagram shows an insect feeding on a plant stem.

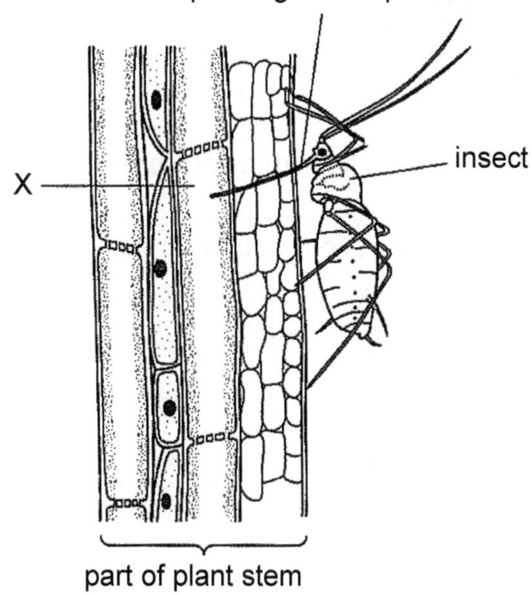

   What is structure X?

   A companion cell
   B mesophyll cell
   C sieve tube element
   D xylem vessel ( )

   [N12/1/12]

3. The diagrams show simple experiments used to demonstrate the processes of respiration and transpiration in plants. All were carried out in the light.

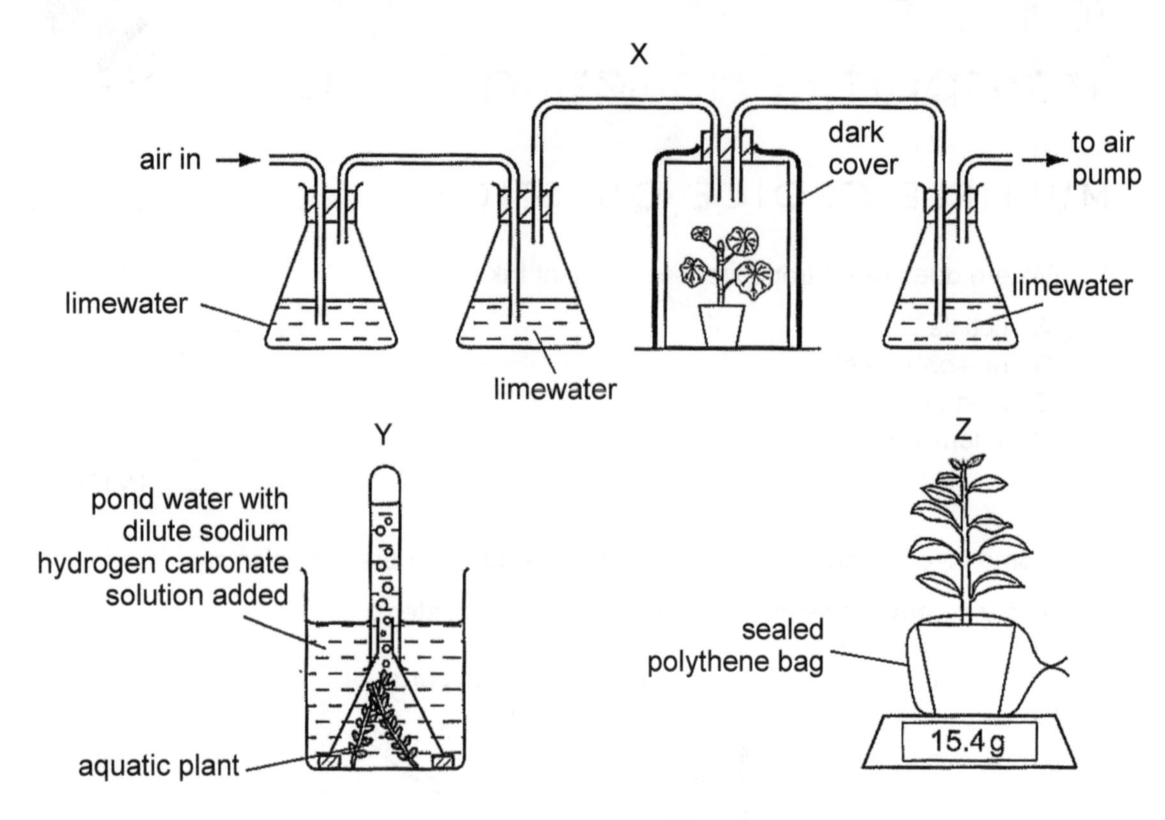

Which experiment is used to demonstrate each process?

	photosynthesis	respiration	transpiration
A	X	Z	Y
B	Y	X	Z
C	Y	Z	X
D	Z	Y	X

( )

[N12/1/13]

4. The diagrams represent some plant cells seen in a section of a stem.

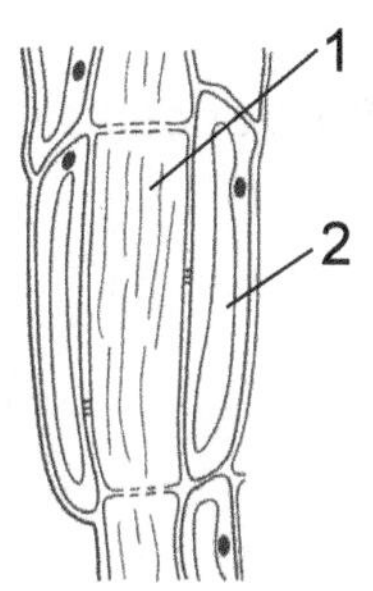

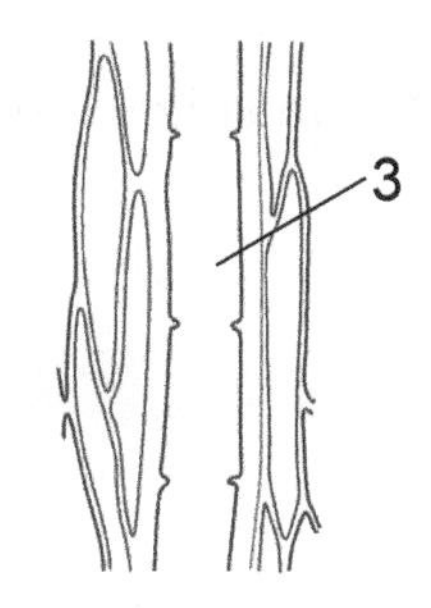

Which cells have the functions shown?

	1	2	3
A	support of young stem	transport of water	transport of sucrose
B	transport of amino acids	supply of energy to surrounding cells	transport of minerals
C	transport of sucrose	transport of water	transport of amino acids
D	transport of water	supply of energy to surrounding cells	support of young stem

( )
[N11/1/11]

5. What is a result of the action of stomata?

	action of stomata	result
A	closed	no water is lost from leaves
B	closed	photosynthesis stops
C	open	net carbon dioxide diffuses in during daylight
D	open	oxygen and water vapour diffuse out at night

( )
[N11/1/12]

6. Which process contributes **most** to the rise of water in the xylem?

A capillarity
B osmosis
C root pressure
D transpiration

( )
[N11/1/13]

7. The cell sap of a root hair has a higher concentration of nitrate ions than the surrounding soil.

Which feature of the cell maintains the higher concentration of these ions in the cell sap?

A cell membrane
B cell wall
C large surface area
D large vacuole

( )
[N10/1/12]

8. The diagram shows a cross-section through a stem. The tissues labelled X in the stem were treated with a poison that prevents respiration. As a result, the movement of sugars through X slowed down.

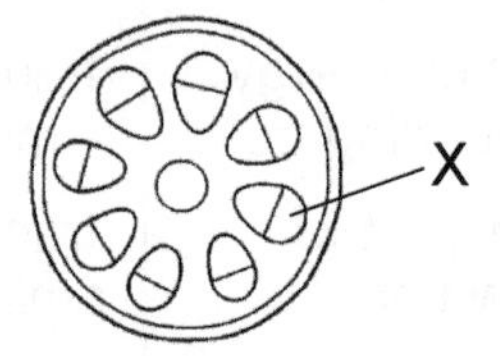

What identifies the tissue and explains the observation?

	tissue	explanation: energy not released by
A	phloem	mitochondria in the companion cells
B	phloem	mitochondria in the sieve tubes
C	xylem	mitochondria in the vessels
D	xylem	nuclei in the vessels

( )
[N10/1/13]

9. A plant stem was dissected into a number of different tissues. Each tissue was tested for the presence of starch, protein and reducing sugar. The results are shown in the table.

Which tissue is xylem?

	starch	protein	sugar
A	✓	×	✓
B	✓	×	×
C	×	✓	✓
D	×	×	×

key
✓ = substance present
× = substance absent

( )
[N09/1/12]

10. The diagram shows a green plant.

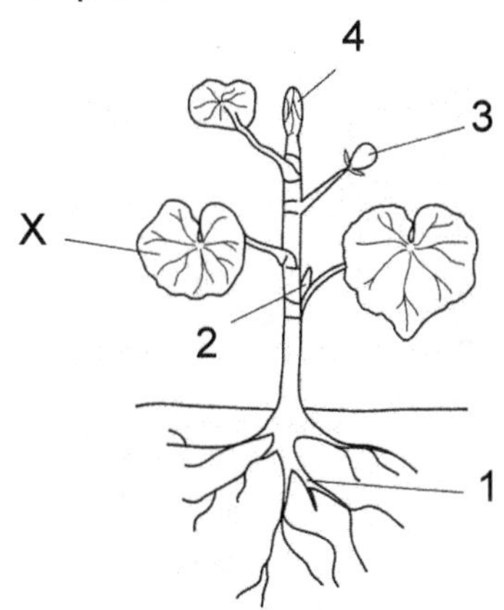

Where will food made by leaf X be found after translocation?

A 1 only
B 1 and 2 only
C 1, 2 and 3 only
D 1, 2, 3 and 4 ( )

[N09/1/13]

11. Transpiration enables water to reach the top of trees.

Which of these statements correctly describes part of this process?

1 Water evaporates into the air spaces of the leaf.
2 Water molecules cohere so are pulled upwards in the xylem.
3 Water passes by osmosis through the stomata.
4 Water vapour diffuses through the mesophyll cells.

A 3 only
B 1 and 2 only
C 2 and 4 only
D 1, 3 and 4 only ( )

[N08/1/11]

12. An experiment was performed on a young plant using an aphid stylet to measure the rate of transport in the phloem.

The same plant was then placed in a bell jar and a chemical placed in the bell jar to absorb all the oxygen present. The rate of transport in the phloem decreased and then stopped.

What is the reason for this?

A Companion cells no longer produce sufficient energy.
B Mitochondria in the xylem vessels cease to function.
C Photosynthesis cannot occur in the bell jar.
D Translocation occurs only by diffusion. ( )

[N08/1/12]

13. Four similar plants are growing under different conditions of temperature and humidity.

Which plant will wilt first?

	temperature	humidity
A	high	high
B	high	low
C	low	high
D	low	low

( )

[N08/1/13]

14. The diagram shows some plant root cells.

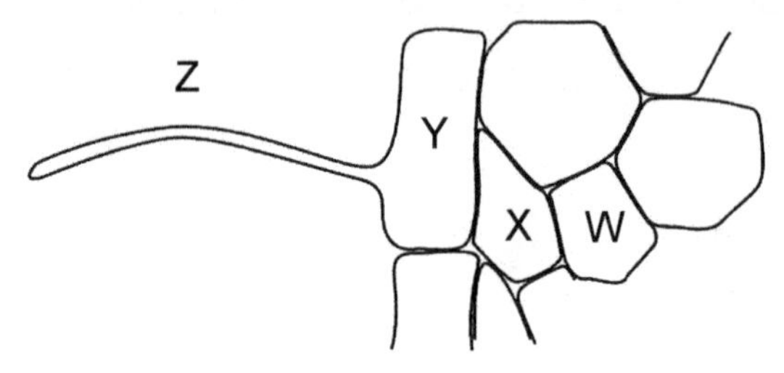

Which statement is correct?

A The water potential of the soil water Z is zero.
B The water potential of cell W is the lowest.
C The water potential of cell X is higher than cell Y.
D The water potential of cell X is lower than cell W.

( )

[N07/1/11]

15. What do phloem and xylem vessels transport in a plant?

	phloem	xylem
A	water	amino acids
B	amino acids	sucrose
C	water	sucrose
D	sucrose	water

( )

[N07/1/12]

16. Which of the following environmental conditions would cause rapid transpiration?

	air	light	temperature
A	damp	bright	cold
B	damp	dim	warm
C	dry	bright	warm
D	dry	dim	cold

( )

[N06/1/12]

17. The photomicrograph shows part of a section of a plant.

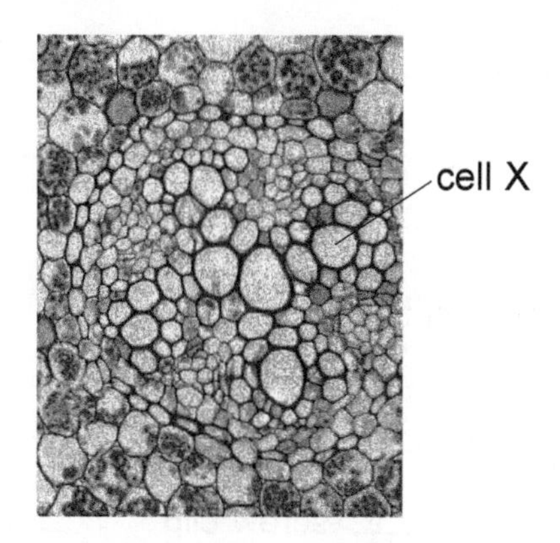

Samples of the contents of cell X were tested.

What results are expected?

	Benedict's reagent	iodine
A	+	+
B	+	–
C	–	+
D	–	–

key
+ = positive result
– = negative result

( )
[N06/1/13]

18. Which conditions cause the fastest rate of transpiration?

A dry and cold
B dry and warm
C wet and cold
D wet and warm

( )
[N05/1/11]

19. The diagram shows a cross-section of a root.

Which tissue transports water to the leaves?

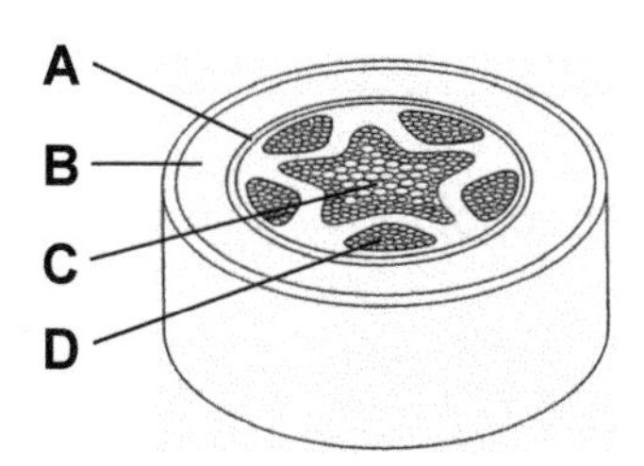

( )
[N05/1/12]

STRUCTURED QUESTIONS . . . .

1. (a) Define the term *transpiration*.

....

....[1]

(b) The figure shows a potometer.

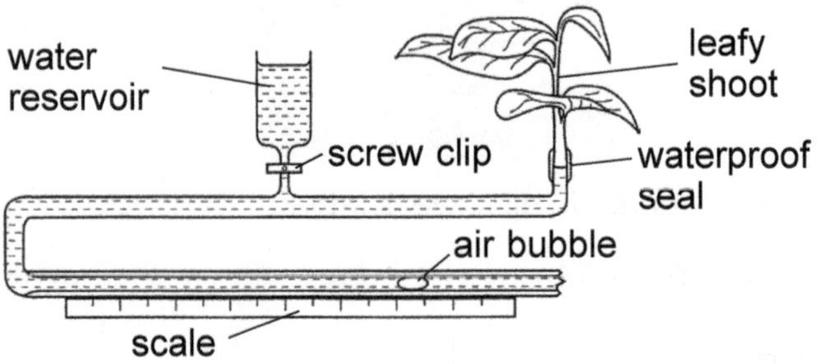

(i) Describe how the potometer can be used to measure the rate of water loss from the leafy shoot.

....

....

....

....[3]

(ii) Explain the purpose of the waterproof seal.

....

....

....[2]

The table shows the results of an investigation, using a potometer, to measure the rate of water loss from a leafy shoot of a plant kept in different conditions.

distance moved by bubble / mm	
still air	moving air
8	16

(c) State and explain the effect of moving air on the water loss from the leafy shoot.

....

....

....

....[3]

[N10/2/7]

2. **Fig. (a)** shows a cross section through a plant stem.

Fig. (b) shows a vertical section through the same stem.

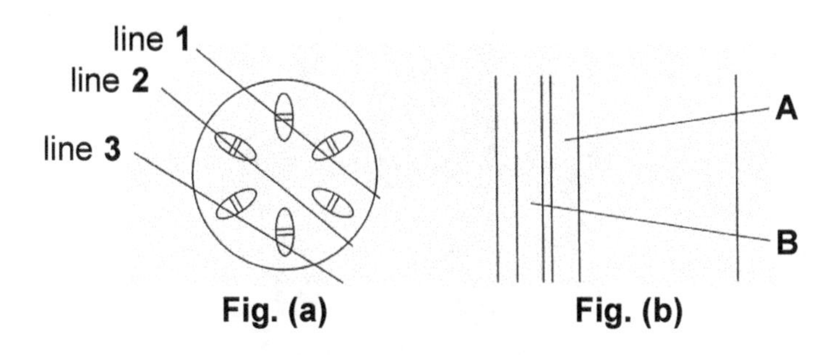

Fig. (a) **Fig. (b)**

(a) Identify the line, 1, 2 or 3, on **Fig. (a)** which represents the section shown in **Fig. (b)**.

....

....[1]

(b) Identify the tissues labelled **A** and **B**.

A ....

....

B ....

....[2]

(c) Name the tissue which transports

(i) food materials, ...

....[1]

(ii) water. ....

....[1]

[N09/2/2]

3. **Fig. (a)** is a transverse section of part of a leaf.

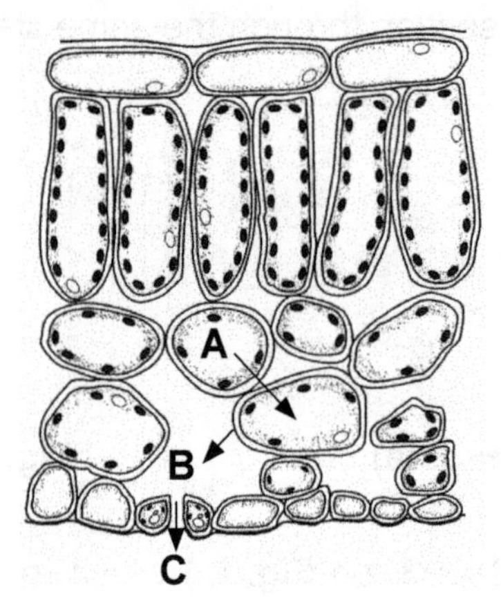

⟶ movement of water

Fig. (a)

(a) Name the processes, represented by the arrows **A**, **B** and **C**, in which water moves.

A ....

....

B ....

....

C ....

....

[3]

(b) State in which tissue, ions such as potassium, move through a plant stem to a leaf.

....

....[1]

(c) Fig. (b) shows the uptake of potassium ions by root hairs when oxygen is present and when oxygen is absent.

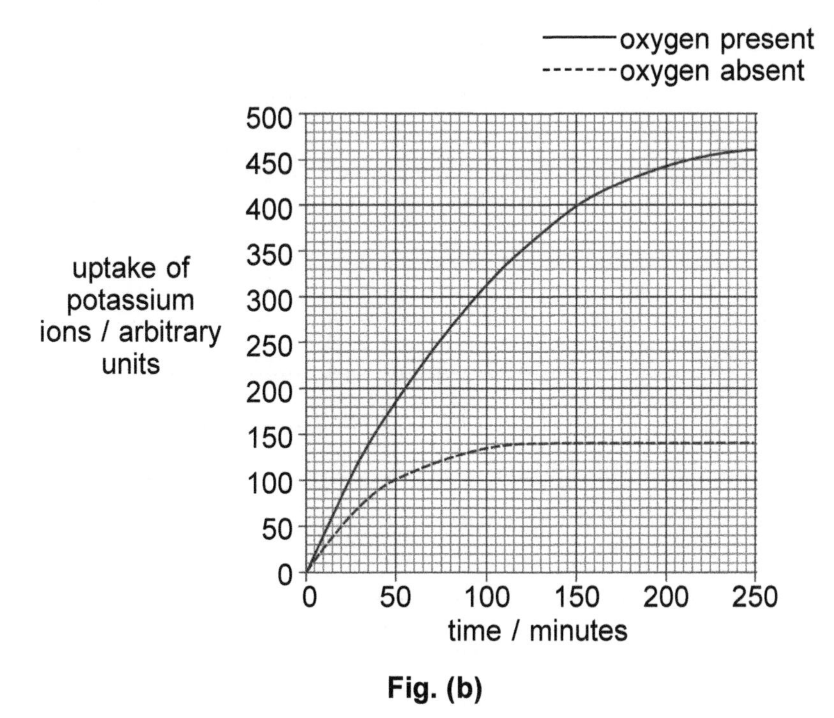

Fig. (b)

(i) State the quantity of potassium ions that has been taken up after 100 minutes in the presence of oxygen.

....

....[1]

(ii) Calculate the difference in the uptake of potassium ions when oxygen is present and when oxygen is absent at 250 minutes.

(You may use the space below for your working.)

....[2]

(d) State three conclusions that may be drawn from the data in the graph.

1. ....

....

....

....

....

2. ....

....

....

....

....

3. ....

....

....

....

....

[3]

[N07/2/4]

4. Two shoots, **J** and **K**, were removed from a different plant and placed in glass jars as shown in **Fig. (a)**.

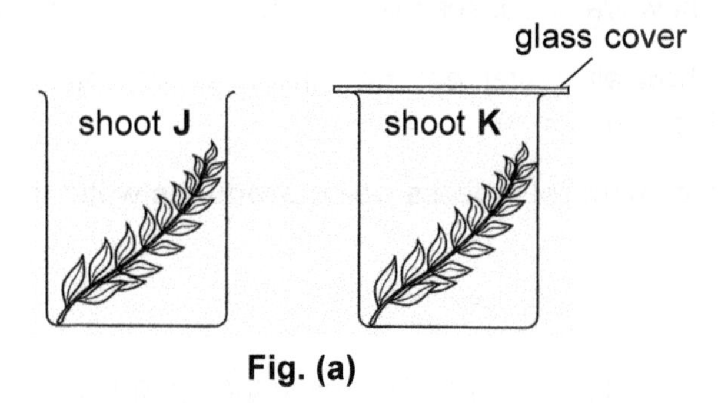

Fig. (a)

Fig. (b) shows the shoots after 12 hours.

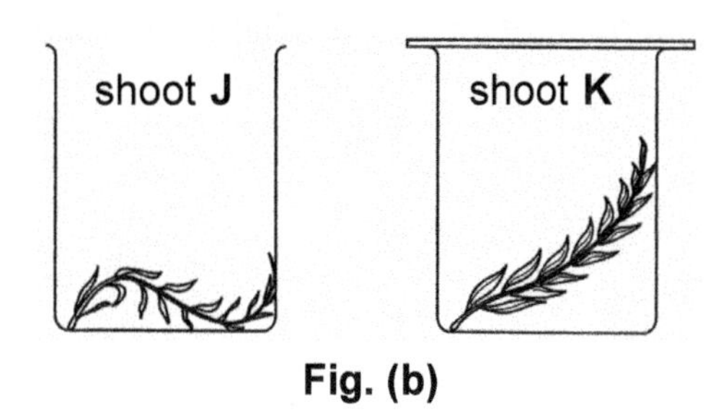

Fig. (b)

(a) State one difference between the air in the jar containing shoot **K** and the air in the jar containing shoot **J** in **Fig. (b)** after 12 hours.

....

....[1]

(b) Explain the difference in the appearance of the two shoots in **Fig. (b)** after 12 hours.

....

....

....

....[3]

[N06/2/5(b)]

FREE RESPONSE QUESTIONS...............

1. **(a)** Explain how water passes from a mesophyll cell to the atmosphere. [5]

   (b) Explain how air movement may affect the loss of water from the surface of a leaf to the atmosphere. [4]

   (c) Name the chemical process which **produces** water inside a mesophyll cell in a leaf. [1]

   [N12/2/9]

2. **(a)** Explain how a plant supports itself in the upright position. [5]

   (b) **(i)** Explain the process of wilting in a plant.

   (ii) Describe the conditions in which wilting is mostlikely to occur. [5]

   [N05/2/7]

TOPIC 7

Transport in Humans

MULTIPLE CHOICE QUESTIONS . . . .

1. Both equations, **Y** and **Z**, show reactions that take place in red blood cells.

Y $Hb + O_2 \leftrightarrow HbO_2$ Hb = haemoglobin

Z $CO_2 + H_2O \leftrightarrow H_2CO_3$

What are the factors that determine the speed and direction of the reactions?

	reaction **Y**		reaction **Z**	
	concentration of reactants	correct enzyme present	concentrations of reactants	correct enzyme present
A	✓	✓	✓	×
B	✓	×	✓	×
C	✓	×	✓	✓
D	×	✓	×	✓

key
✓ = factor involved
× = factor not involved

( )
[N12/1/14]
[N08/1/17]

2. The diagrams represent the circulatory system.
Which diagram is correct for an adult human?

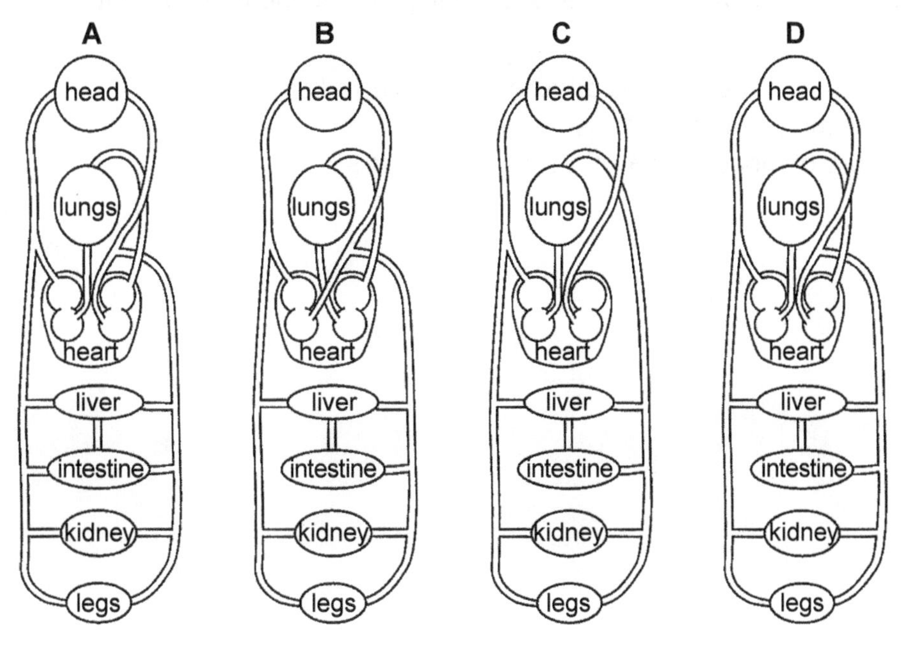

(D)
[N12/1/15]

3. The graph shows the pressure changes in the aorta and the left side of the heart during one cardiac cycle.

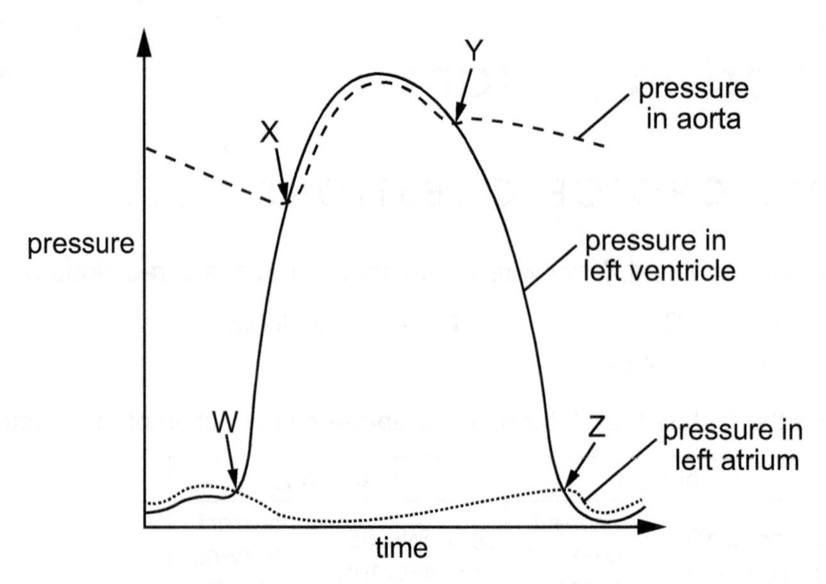

At which points does the aortic valve open and close?

	open	close
A	W	Z
B	X	Y
C	Y	Z
D	Z	W

( )

[N12/1/16]

4. During the cardiac cycle, what causes the flaps of the bicuspid valve to close?

A a nervous impulse to the valve flaps
B contraction of the tendons attached to the valve flaps
C contraction of the valve flaps
D pressure of blood as the left ventricle contracts

( )

[N11/1/15]

5. The arrows on the diagram show the direction of movement of some of the liquids in and around a capillary.

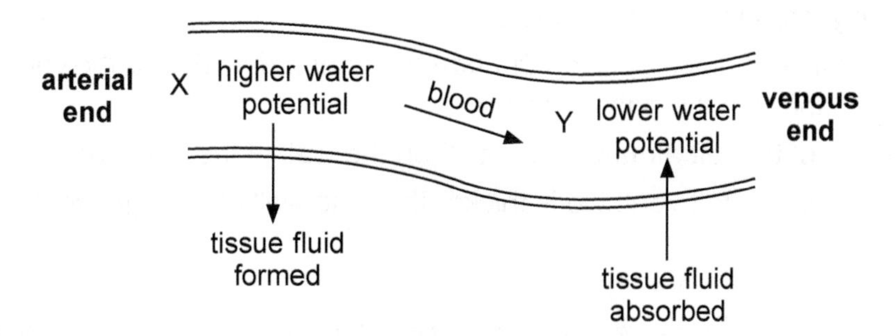

What causes the change in water potential between points X and Y?

A Blood cells stay in the capillary.
B Blood pressure falls between X and Y.
C Oxygen moves out of the capillary at the arterial end.
D Plasma proteins stay in the capillary. ( )

[N11/1/16]

6. The diagram represents part of the circulatory system.

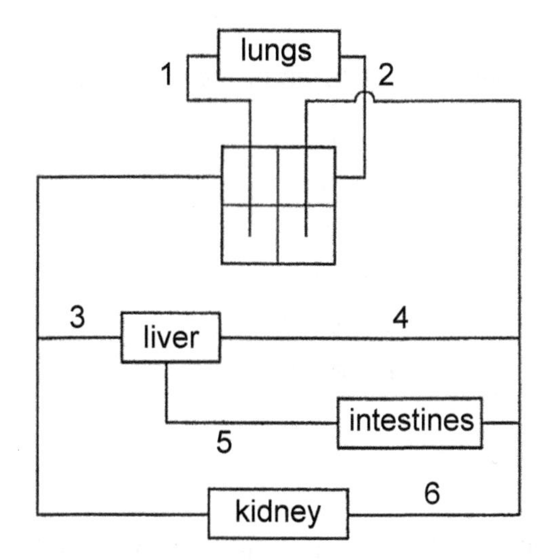

Which row names the blood vessels correctly?

	pulmonary artery	hepatic artery	hepatic vein
A	1	3	4
B	1	4	3
C	2	6	5
D	6	2	5

( )

[N10/1/14]

7. Which statement about the heart is **not** correct?

A The volume of blood passing through the left atrium equals the volume of blood passing through the left ventricle.

B The volume of blood passing through the right atrium equals the volume of blood passing through the right ventricle.

C The wall of the left atrium is thicker than the wall of the left ventricle.

D The wall of the left ventricle is thicker than the wall of the right ventricle. ( )

[N10/1/15]

8. The arrows on the diagram show the direction of movement of some of the substances carried in plasma as they enter and leave a capillary.

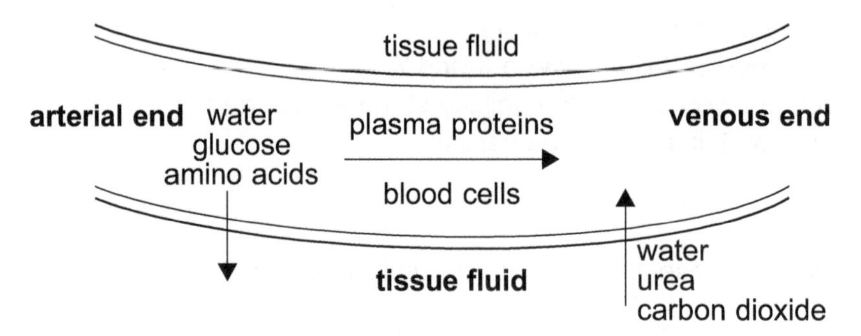

Which two factors cause water to re-enter the capillary at the venous end?

	plasma water potential	blood pressure in the capillary
A	decreased	decreased
B	decreased	increased
C	increased	decreased
D	increased	increased

( )

[N10/1/16]
[N07/1/15]

9. Which substance will pass from muscle cells into the capillary via the tissue fluid?

A adrenaline
B carbon dioxide
C glycogen
D urea ( )

[N09/1/14]

10. Which two changes in arteries would affect the blood pressure as stated?

	circular muscle in wall	lumen size	effect on blood pressure
A	contracts	decreases	increases
B	contracts	increases	increases
C	relaxes	decreases	decreases
D	relaxes	increases	increases

( )
[N09/1/15]

11. The diagram shows the pressure changes in the left side of the heart.

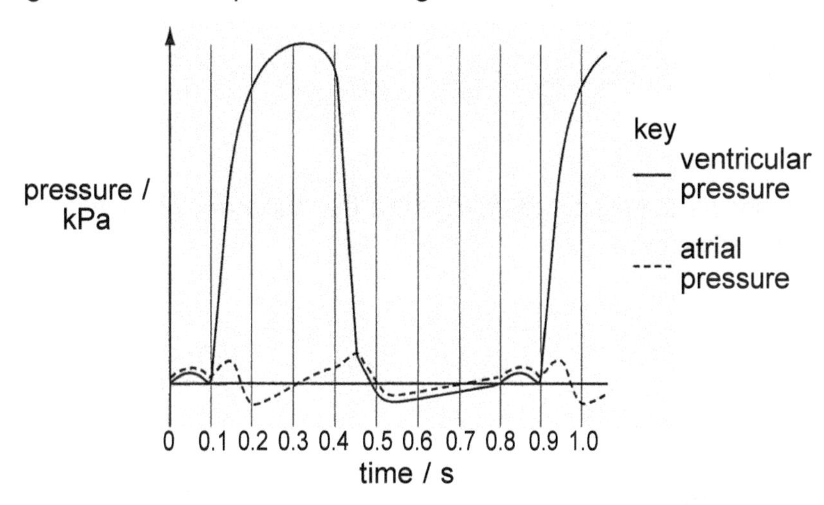

What is the ratio of the timing of atrial systole to atrial diastole?

A 0.1 : 0.3
B 0.1 : 0.7
C 0.1 : 0.8
D 0.7 : 0.1

( )
[N09/1/16]

12. Coronary heart disease is caused by partial occlusion (blockage) of the coronary arteries.

Which blood flow does this reduce?

A into the left atrium
B into the right ventricle
C to the cardiac muscle
D to the intercostal muscles

( )
[N08/1/14]

13. The table shows the blood groups of four people and the type of blood each received in a transfusion.

	blood group under ABO system	blood type received in transfusion
W	A	O
X	B	AB
Y	AB	O
Z	O	AB

Which two people are at risk from agglutination?

A W and X
B W and Y
C X and Z
D Y and Z

( )
[N08/1/15]

14. The graph shows the pressure changes in the aorta and the left side of the heart during one cardiac cycle.

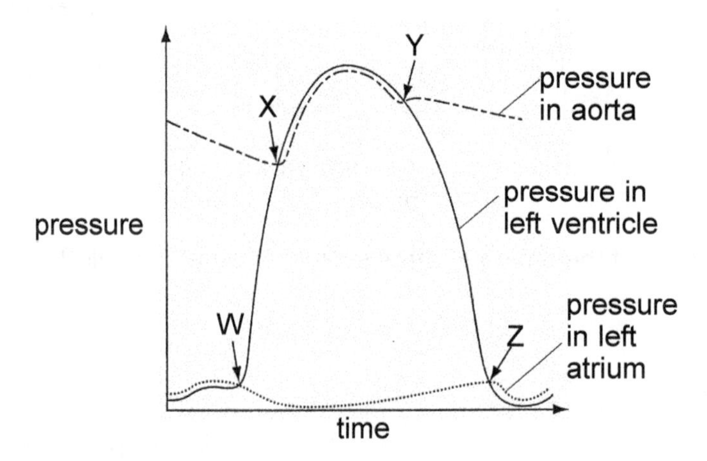

At which points does the bicuspid valve open and close?

	open	close
A	W	Z
B	X	Y
C	Y	Z
D	Z	W

( )
[N08/1/16]

15. Some functions of the blood are listed.

1 antibody formation
2 clotting
3 distribution of hormones
4 phagocytosis

Which two functions are responsible for tissue rejection following a transplant operation?

A 1 and 2
B 1 and 4
C 2 and 3
D 3 and 4

( )
[N07/1/13]

16. Blood samples from three veins in the body was tested for the concentration of oxygen, carbon dioxide and urea. The results, in arbitrary units, are shown in the table.

vein	oxygen concentration	carbon dioxide concentration	urea concentration
1	40	48	1.5
2	40	48	7.5
3	90	40	4.0

The blood from which veins was sampled?

	hepatic vein	pulmonary vein	renal vein
A	1	2	3
B	2	3	1
C	3	1	2
D	3	2	1

( )
[N07/1/14]

17. The diagrams represent the circulatory system.

Which diagram is correct for an adult human?

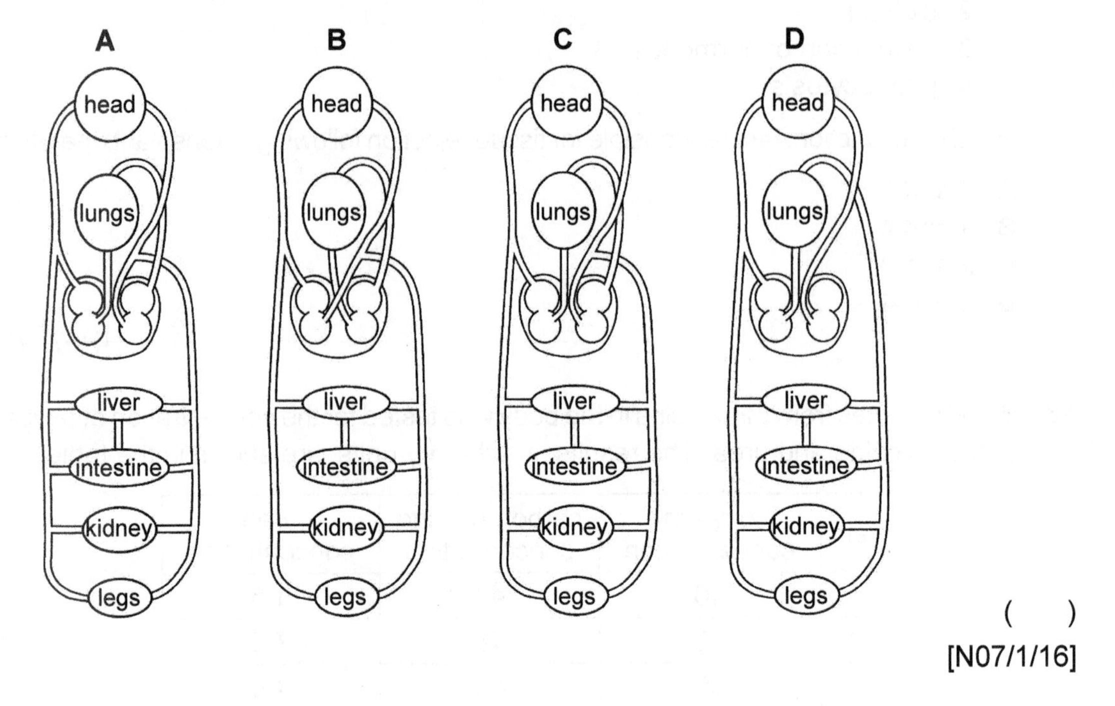

( )

[N07/1/16]

18. The graph shows pressure changes in the left side of the heart, during a single heart beat.

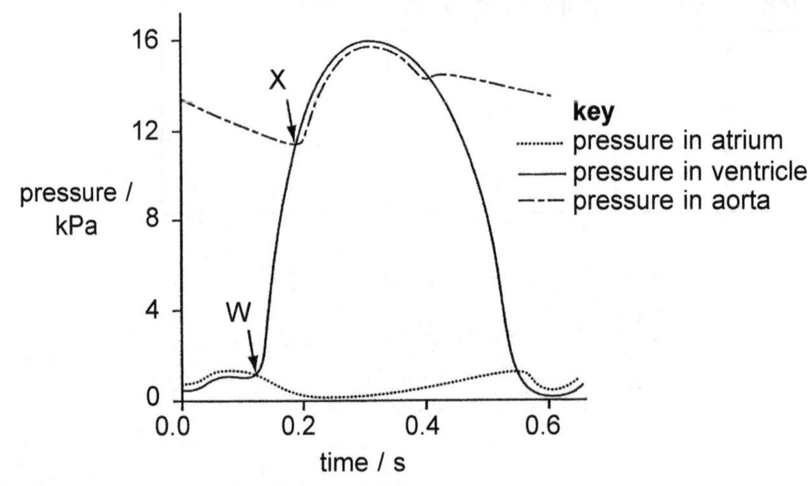

Between points W and X, are the following valves open or closed?

	atrio-ventricular	semi-lunar
A	closed	closed
B	closed	open
C	open	closed
D	open	open

( )

[N06/1/14]

19. The graph shows pressure changes in the left side of the heart, while at rest, during a single heart beat.

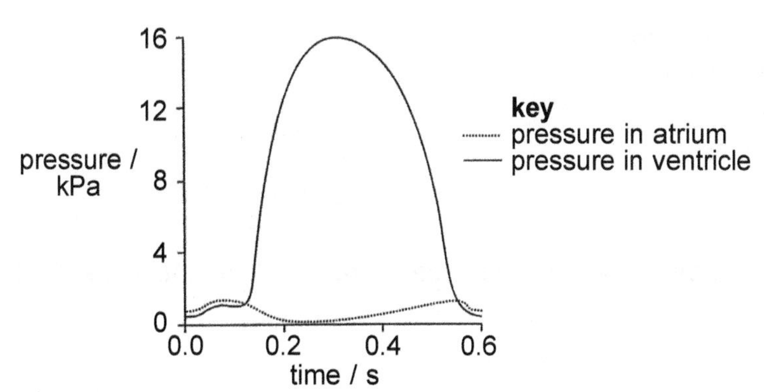

What is the number of times that this person's heart beats in one minute, while at rest?

A 60
B 70
C 100
D 120 ( )

[N06/1/15]

20. Which statement explains why humans are said to have a double circulation?

A As blood circulates it passes twice through the heart.
B Each side of the heart has two chambers.
C Each side of the heart has two valves.
D There are two different sets of arteries leaving the heart. ( )

[N06/1/16]

21. The diagram shows a valve in a section through a blood vessel.

Which statement is correct?

A Blood flows from **X** to **Y**, opening the valve.
B Muscles in the wall contract and close the valve, preventing backflow.
C The elastic wall causes the valve to close between heartbeats.
D The valve is forced open when the blood pressure at **Y** is greater than at **X**.

( )

[N05/1/13]

22. Which blood vessel carries blood with the **lowest** concentration of urea?

A hepatic portal vein
B pulmonary vein
C renal vein
D vena cava

( )
[N05/1/14]

23. Which substance passes out of the capillaries and into the tissue fluid around muscle cells?

A carbon dioxide
B glucose
C glycogen
D urea

( )
[N05/1/15]

24. When the skin is cut the blood clots.

In which order would the components of the blood become involved?

	first	→		last
A	fibrin	platelets	red blood cells	fibrinogen
B	fibrinogen	red blood cells	platelets	fibrin
C	platelets	fibrin	fibrinogen	red blood cells
D	platelets	fibrinogen	fibrin	red blood cells

( )
[N05/1/16]

STRUCTURED QUESTIONS . . . .

1. **(a)** The figure shows the relationship between a blood capillary and a body tissue.

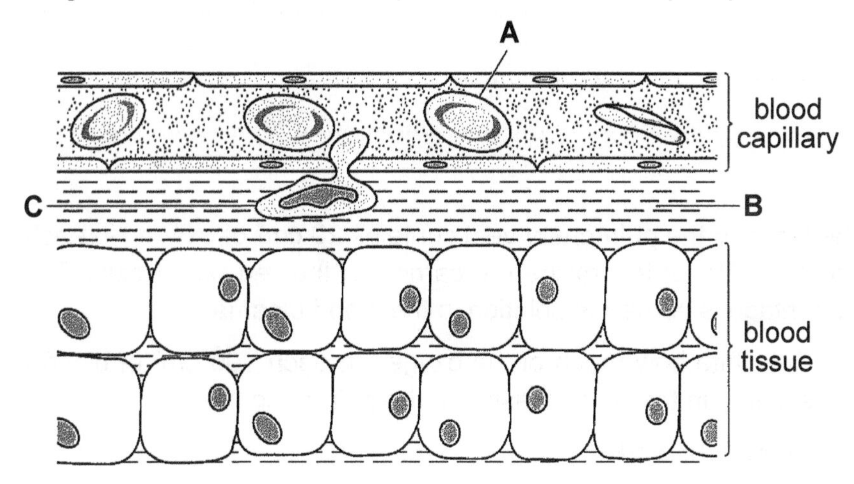

(i) Name the parts labelled **A** and **B**.

A...

B...[2]

(ii) Draw an arrow on the figure to show the movement of oxygen into body cells. [1]

(iii) State two functions of the cell labelled **C**.

1. ....

....

2. ....

....[2]

(b) The table gives information about blood transfusions.

donor	recipient			
	group A	group B	group AB	group O
group A			✓	×
group B	×	✓	✓	×
group AB	×	×	✓	
group O	✓	✓	✓	✓

Key

✓ indicates successful transfusion is possible

× indicates successful transfusion is not possible

Complete the table for donor group A and for donor group AB. [3]

[N12/2/5]

2. **(a)** State and explain two ways in which red blood cells are adapted to their function.

1. ....

....

2. ....

....

[2]

(b) The blood in the pulmonary vein contains 20 cm^3 of oxygen in each 100 cm^3 of blood. 98.7% of the oxygen is carried in the red blood cells. The remaining percentage is carried in solution in the blood plasma.

(i) Calculate how much of the oxygen in each 100 cm^3 of blood is carried in solution in the blood plasma in the pulmonary vein.

Show your working.

.... [2]

(ii) State and explain the effect on the concentration of oxygen carried in the red blood cells of breathing in air containing tobacco smoke.

....

....

....

....

....[3]

(c) Describe the role of white blood cells in protecting the body from disease.

....

....

....

....[2]

[N11/2/3]

3. The figure shows an external view of the human heart.

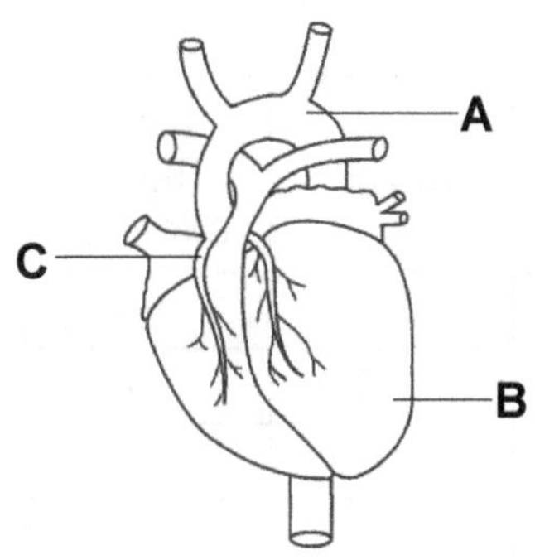

(a) Identify the structures labelled **A** and **B**.

A ....

....

B ....

....[2]

(b) State the function of the blood vessel labelled **C**.

....

....

....

....[2]

(c) **(i)** At rest the rate of blood flow to the heart muscle is 260 cm^3 per minute. During exercise this rate rises to 780 cm^3 per minute.

Calculate the percentage increase in the rate of blood flow during exercise.

Show your working.

....% [2]

(ii) Explain the advantage of this increase in rate of blood flow.

....

....

....[2]

[N10/2/1]

4. The table shows the relative volumes of blood flowing through different organs at rest and during exercise.

organ	at rest	during light exercise	during heavy exercise
brain	1.0	1.1	1.1
skin	1.0	2.4	3.1
kidneys	1.0	0.8	0.5
muscles	1.0	3.8	11.0
heart	1.0	1.6	2.5

(a) State which organ has the greatest increase in blood supply.

....

....[1]

(b) Explain the changes in blood volume for

(i) the muscles, ...

....

....[3]

(ii) the skin. ....

....

....[2]

[N09/2/3]

5. The figure shows a section through the heart.

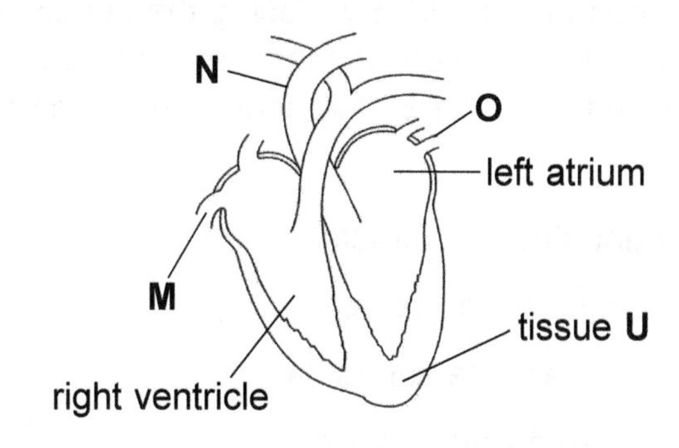

(a) **(i)** Identify blood vessels **M**, **N** and **O** in the figure.

M ....

....

N ....

....

O ....

....[3]

(ii) State the purpose of tissue **U** in the functioning of the heart.

....

....[1]

(iii) Name the blood vessel which, unless blocked by heart disease, would normally supply this tissue with oxygen.

....

....[1]

(b) On the figure, draw, in the correct position, and label

(i) the semilunar valves,

(ii) the tricuspid valve,

(iii) the bicuspid (mitral) valve. [4]

(c) The four statements below describe some of the events that occur during the flow of blood through the heart. By placing the numbers **1** to **4** in the boxes, indicate the correct sequence of these events, starting immediately after deoxygenated blood has entered the heart and ending as the blood is sent to the lungs.

☐ The right atrium contracts.

☐ The semilunar valves open.

☐ The right ventricle contracts.

☐ The tricuspid valve closes.

[3]

[N05/2/5]

FREE RESPONSE QUESTIONS. . . . .

1. (a) The figure shows the mean heart rate of groups of rats at different ages. Each group was given a different treatment each day.

- Group **A** was given electric shocks to the heart muscle.
- Group **B** was injected with a drug which blocks nerve impulses to the heart muscle.

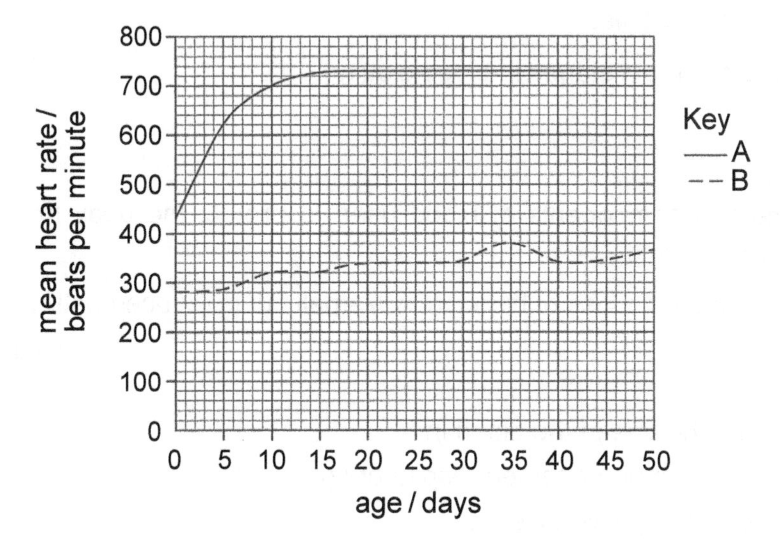

(i) The table shows the mean heart rate for another group of rats, Group **C**, which was not given any treatment.

	age / days										
	0	5	10	15	20	25	30	35	40	45	50
mean heart rate/ beats per minute	300	390	430	460	480	500	510	520	490	470	460

Plot the data for Group **C** on the figure. [3]

(ii) State the difference in the mean heart rates of the rats in Group **A** and those in Group **B** at day 25. [1]

(iii) State two differences between the mean heart rate of the rats in Group **C** and those in Group **B**. [2]

(b) Explain how the drug given to the rats in Group **B** was transported to the heart muscle. [4]

[N11/2/8]

2. The table shows the effect of temperature on the clotting time of the blood.

temperature / °C	10	15	20	25	30	35	40	45
clotting time / s	86	58	48	40	30	24	32	58

(a) Plot a graph of these data. [5]

(b) With reference to the table, describe the relationship between temperature and blood clotting time. [3]

(c) Outline the main stages in blood clotting. [4]

[N09/2/8]

3. (a) Name two components of blood used to protect the body from the entry of microorganisms. [2]

(b) Explain how the loss of blood from a small cut is reduced naturally. [3]

[N08/2/10(a)(i)(ii)]

4. (a) Describe the role of the blood in

(i) the transport of named respiratory gases,

(ii) defence against disease. [7]

(b) Outline the link between coronary heart disease and diet. [3]

[N06/2/6]

TOPIC 8

Respiration in Humans

MULTIPLE CHOICE QUESTIONS . . . .

1. What happens to the volume of the lungs and the air pressure within them during inspiration?

	lung volume	air pressure
A	decreases	decreases
B	decreases	increases
C	increases	decreases
D	increases	increases

( )
[N12/1/17]

2. Carbon dioxide turns limewater cloudy (milky).

Which of the following would demonstrate that expired air contains more carbon dioxide than inspired air?

A

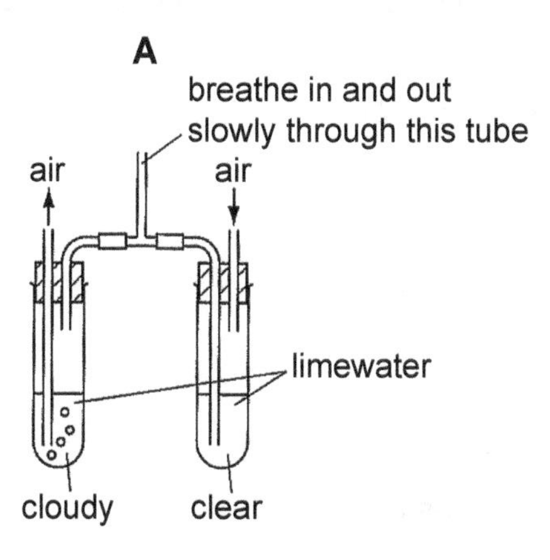

B

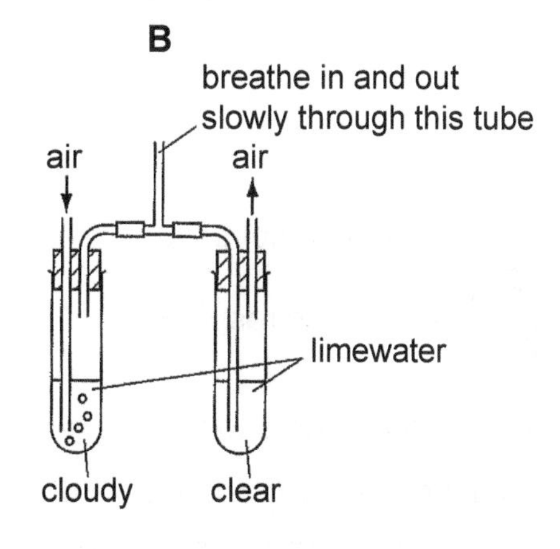

C

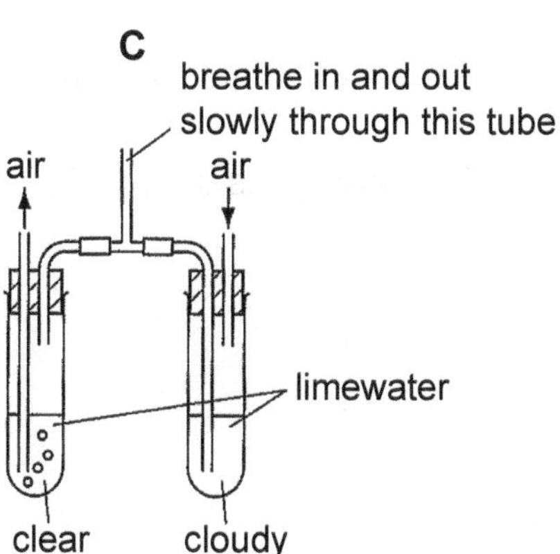

D

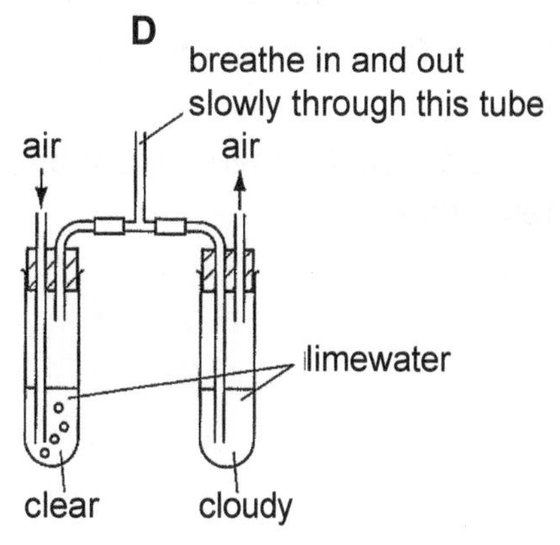

( )
[N12/1/18]

3. The bar chart shows the percentage of energy released by aerobic and anaerobic respiration during races run by athletes over different distances.

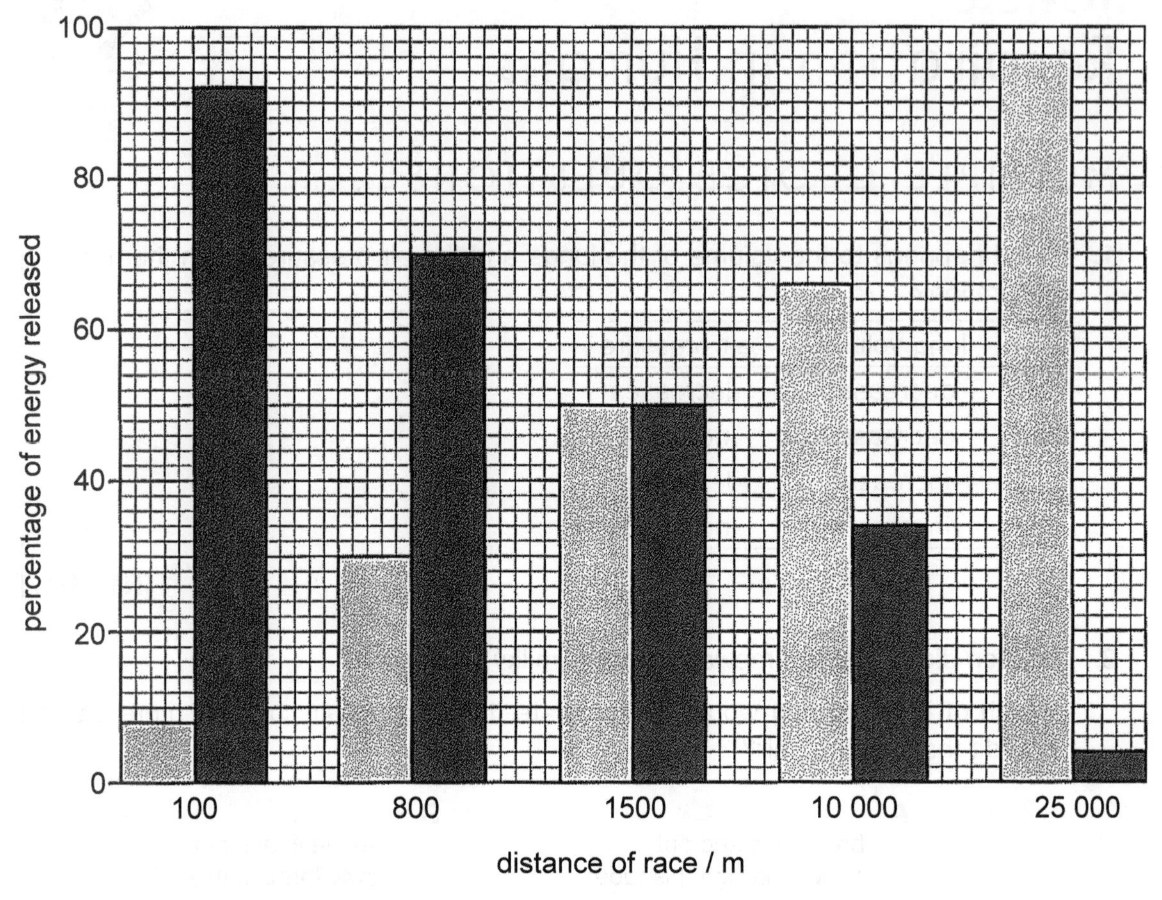

key

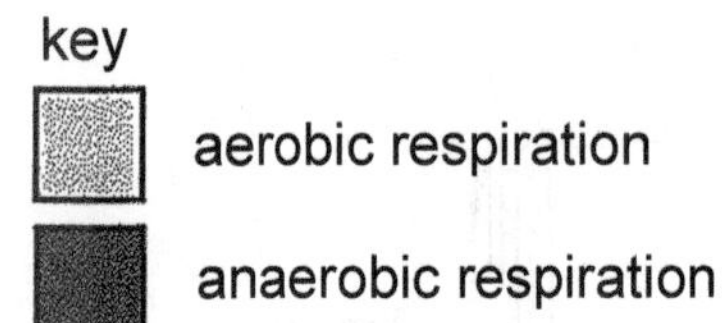

What does the chart show?

A An athlete running a distance of 1500 metres does not build up an oxygen debt.

B At rest, no anaerobic respiration occurs.

C In a 25 000 metre event, an athlete will not be capable of sprinting at the end of the race.

D In races shorter than 1500 metres, most of the energy is supplied by anaerobic respiration.

( )

[N12/1/19]

4. The diagram shows a red blood cell and some of the reactions that occur in it.

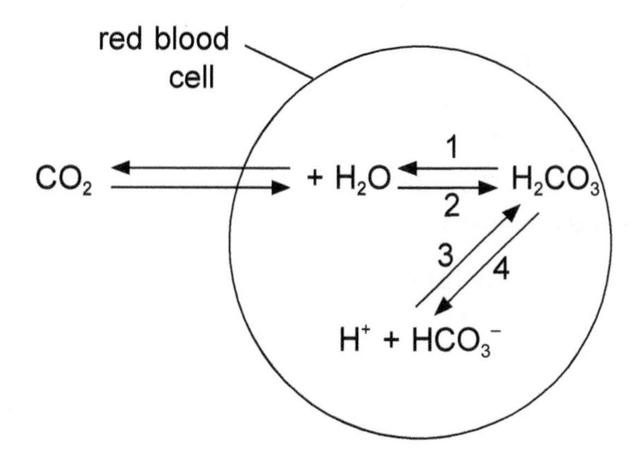

Which numbered reaction(s) involve carbonic anhydrase?

A 1 and 2
B 1 only
C 3 and 4
D 4 only

( )

[N11/1/17]

5. In vigorous exercise, which row gives the correct state of the respiratory muscles during expiration?

	diaphragm muscles	internal intercostal muscles	external intercostal muscles
A	contracted	contracted	relaxed
B	contracted	relaxed	contracted
C	relaxed	contracted	relaxed
D	relaxed	relaxed	contracted

( )

[N11/1/18]

6. The table compares aerobic and anaerobic respiration in muscle cells.

Which comparison is correct?

	level of energy released		level of fatigue produced	
	aerobic respiration	anaerobic respiration	aerobic respiration	anaerobic respiration
A	+	–	+	–
B	+	–	–	+
C	–	+	+	–
D	–	+	–	+

key
+ = more
– = less

( )

[N11/1/19]

7. Some effects of smoking tobacco are listed.

1 bronchitis
2 increase in alertness
3 increase in blood pressure
4 increase in heart rate
5 increase in mucus production
6 uncontrolled cell division

Which effects are caused by tar?

A 1, 2 and 3
B 1, 5 and 6
C 2, 4 and 6
D 3, 4 and 5 ( )

[N10/1/17]

8. The pulse rate of a girl was measured every two minutes and plotted on the graph.

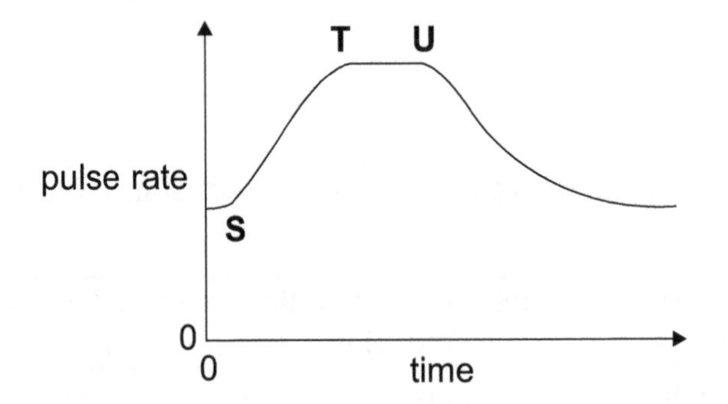

Her exercise started at **S** and finished at **T** but her pulse rate did not start to drop until **U**.

Which process(es) would occur during the **T-U** interval?

1 accumulation of lactic acid from muscle cells
2 increased supply of oxygen to the muscle cells
3 increased transport of carbon dioxide to the lungs

A 1, 2 and 3
B 1 and 3 only
C 2 only
D 2 and 3 only ( )

[N10/1/18]

9. The graph shows changes in air pressure in the lungs during breathing.

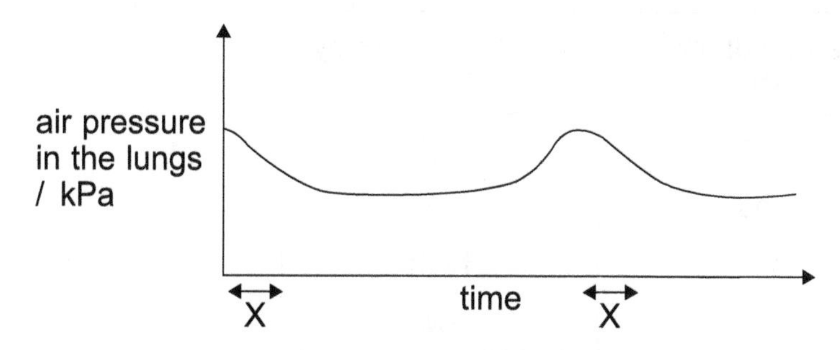

What causes the change in air pressure during period **X**?

A contraction of the diaphragm muscles
B decrease in the volume of the lungs
C movement of ribs downwards
D relaxation of the external intercostal muscles ( )

[N10/1/19]

10. Which reaction is catalysed by carbonic anhydrase when red blood cells pass through the lungs?

A $CO_2 + H_2O \rightarrow H_2CO_3$
B $HCO_3^- + H^+ \rightarrow H_2CO_3$
C $H_2CO_3 \rightarrow CO_2 + H_2O$
D $H_2CO_3 \rightarrow H^+ + HCO_3^-$ ( )

[N09/1/17]

11. Some effects of smoking are listed.

1 bronchitis
2 uncontrolled division in some cells
3 increase in alertness
4 increase in heart rate
5 increase in mucus production
6 increase in blood pressure

Which effects are caused by nicotine?

A 1, 2 and 3
B 1, 2 and 5
C 3, 4 and 6
D 4, 5 and 6 ( )

[N09/1/18]

12. What is correct for anaerobic respiration in muscles?

	carbon dioxide produced	amount of energy released
A	no	high
B	no	low
C	yes	high
D	yes	low

( )

[N09/1/19]

13. The diagram shows part of the lining of the human trachea.

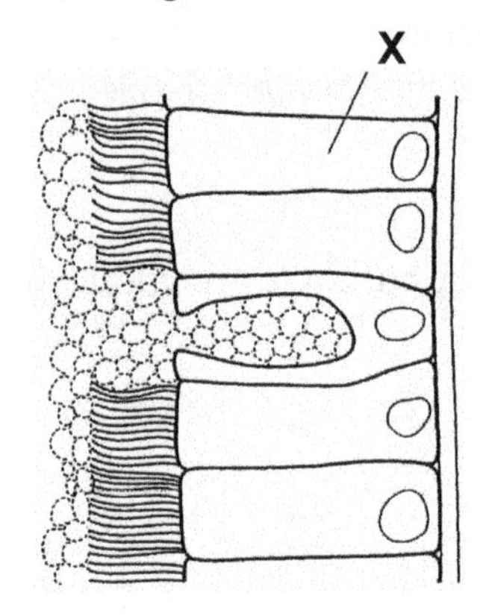

What is the function of cell **X**?

A gaseous exchange
B mucus removal
C phagocytosis
D secretion of mucus

( )

[N08/1/18]
[N05/1/18]

14. What happens during the process of breathing in?

	external intercostal muscles	internal intercostal muscles	diaphragm
A	contract	relax	contracts
B	contract	relax	relaxes
C	relax	contract	contracts
D	relax	contract	relaxes

( )

[N08/1/19]

15. Carbon dioxide turns limewater cloudy (milky).

Which one of the following would demonstrate that expired air contains much carbon dioxide?

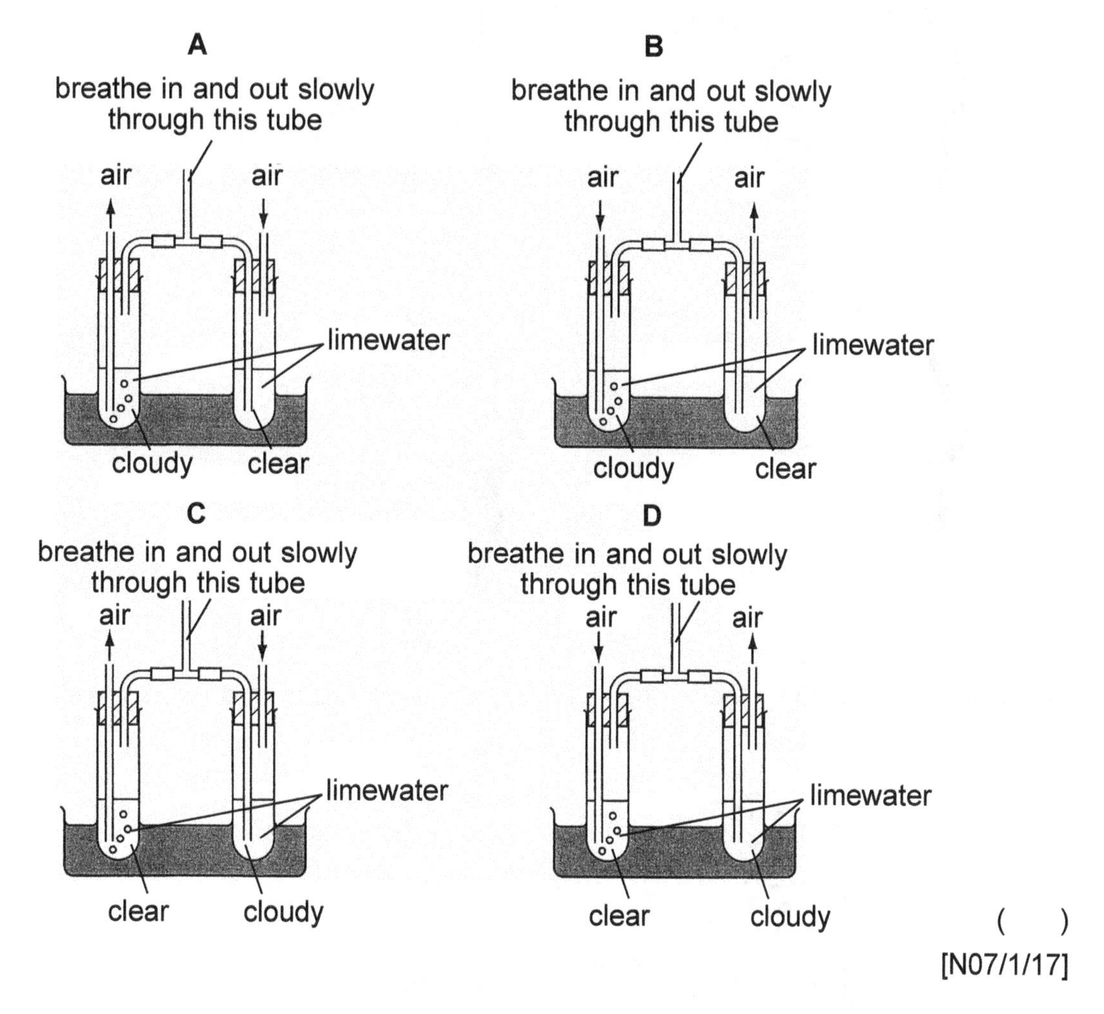

( )

[N07/1/17]

16. As a person breathes in, what happens to the diaphragm and to the rib cage?

	diaphragm becomes	rib cage moves
A	flatter	downwards and inwards
B	flatter	outwards and upwards
C	curved	downwards and inwards
D	curved	outwards and upwards

( )

[N07/1/18]

17. Which diagram shows the most efficient respiratory surface?

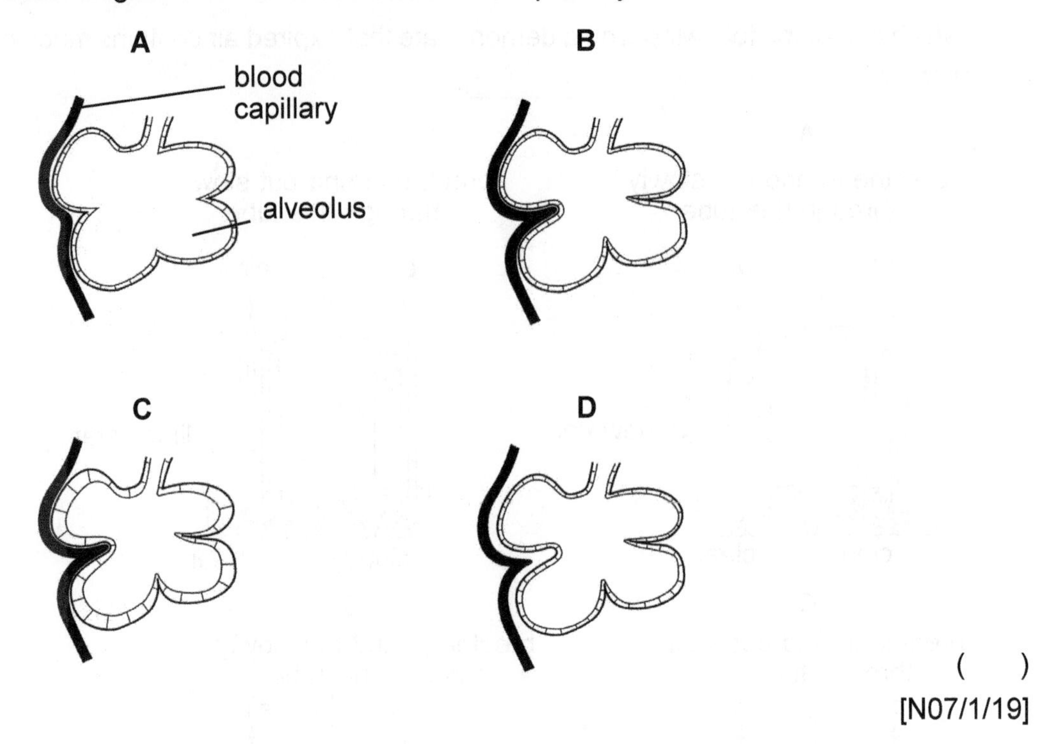

( )

[N07/1/19]

18. The diagrams show the epithelium lining the bronchioles in a non-smoker and a smoker.

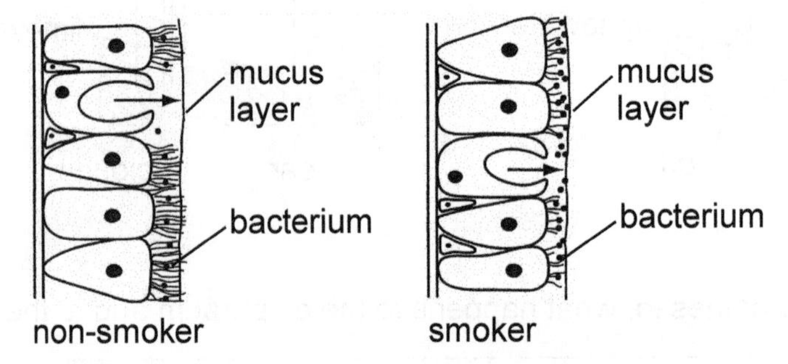

As a result of the changes, what will the smoker experience?

A more lung infections
B more mucus running down the nose
C the bronchioles become wider
D the cilia will beat more rapidly

( )

[N07/1/24]

19. The diagram shows the ribs and some of the muscles used in breathing.

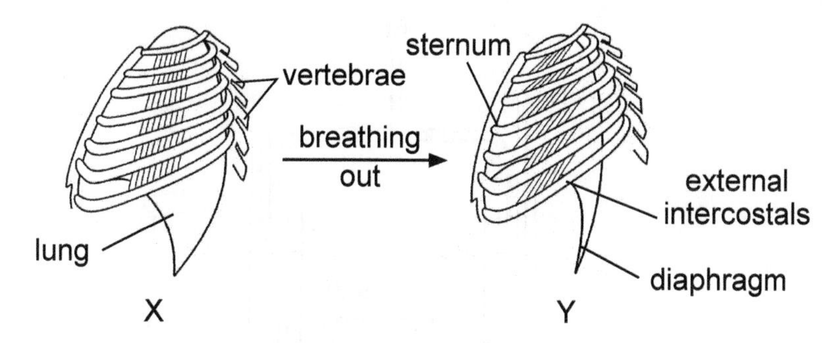

Which muscles relax in moving from position X to position Y?

	diaphragm	external intercostals
A	no	no
B	no	yes
C	yes	no
D	yes	yes

( )

[N06/1/17]

20. Tar and carbon monoxide are present in tobacco smoke.
What are their effects on health?

	tar	carbon monoxide
A	causes high blood pressure	damages haemoglobin
B	causes high blood pressure	is addictive
C	causes lung cancer	damages haemoglobin
D	causes lung cancer	is addictive

( )

[N06/1/26]

21. Which muscle actions occur during inspiration?

	diaphragm muscles	external intercostal muscles
A	contract	contract
B	contract	relax
C	relax	contract
D	relax	relax

( )

[N05/1/17]

22. The apparatus shown is used to investigate gas exchange during breathing.

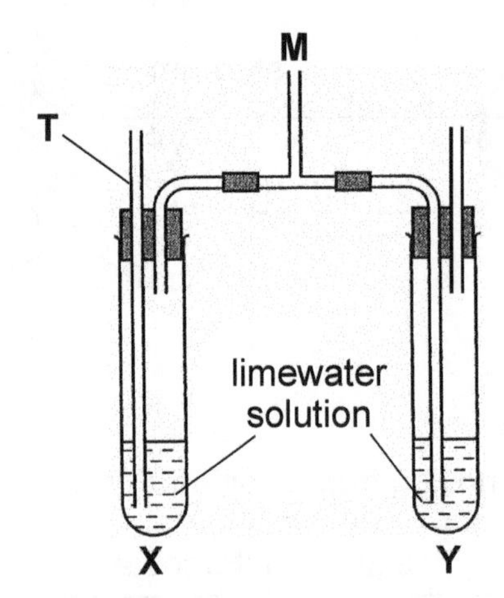

What would occur when a person breaths out through tube **M**?

A The solutions in **X** and **Y** both turn cloudy.
B The solution in **X** remains clear, but that in **Y** turns cloudy.
C The solution in **X** turns cloudy, but that in **Y** remains clear.
D The solution in **X** is forced out through the tube **T**.

( )
[N05/1/20]

23. The table shows results from a study into the effects of smoking while pregnant.

number of cigarettes smoked per day by mother while pregnant	average birth weight of baby / kg	average height of child at 15 years / cm
0	3.7	166.1
1 – 9	3.5	165.0
10 and over	3.1	162.8

Which of these effects of smoking while pregnant is supported by the information in the table?

A increased growth rate and increased birth weight
B increased growth rate and reduced birth weight
C reduced growth rate and increased birth weight
D reduced growth rate and reduced birth weight

( )
[N05/1/24]

STRUCTURED QUESTIONS . . . .

1. **(a)** The table shows the effect of breathing air containing different concentrations of carbon dioxide.

percentage of carbon dioxide in inhaled air / %	volume of each breath / cm^3	breathing rate / breaths per minute
0.03	520	14
1.00	750	16
3.00	1200	18
5.00	2200	25

(i) With reference to the table, state two effects of breathing air containing different concentrations of carbon dioxide.

1. ....

....

2. ....

....

[2]

(ii) Calculate the total volume of air entering the lungs per minute when breathing air containing 3% carbon dioxide.

Show your working.

.... cm^3 [2]

The figure shows a section of an alveolus.

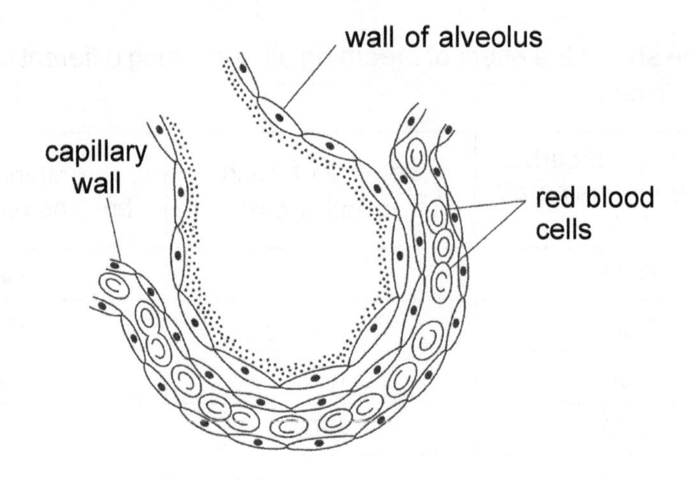

(b) State the function of alveoli and explain how they are adapted to their function.

function ....

....

....

adaptations ....

....

....

....

....

....

....

....

[4]

[N10/2/5]

2. **Fig. (a)** is a diagram of a section of the thorax to show the lungs.

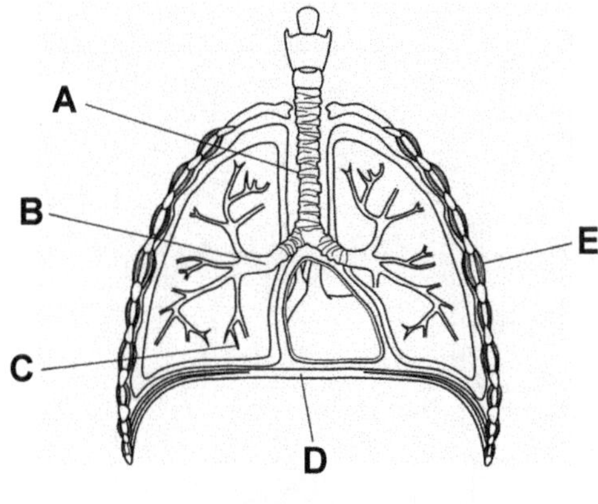

Fig. (a)

(a) Name the structures labelled **A**, **B** and **C**.

A ....

....

B ....

....

C ....

....

[3]

(b) Explain the role of the structures **D** and **E** when breathing in.

....

....

....

....

....

....[4]

Fig. (b) shows the effect that smoking cigarettes has on the pulse rate.

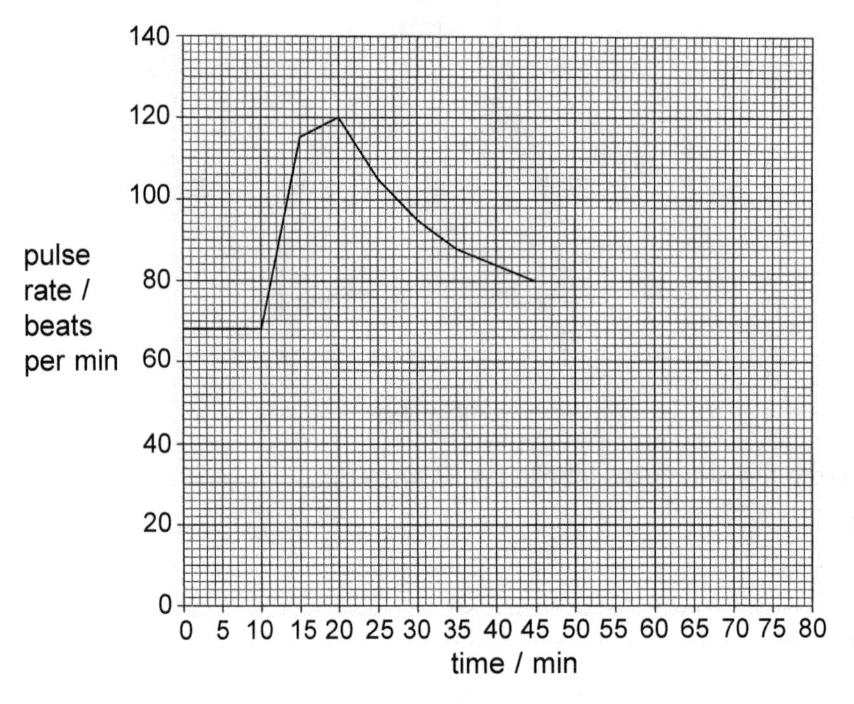

Fig. (b)

Smoking started at the 10th minute and ended at the 20th minute.

(c) **(i)** State the maximum increase in the pulse rate.

....[1]

(ii) Use the graph to determine the time at which the pulse would return to the resting rate.

....[1]

(d) Suggest how smoking can be harmful during pregnancy.

....

....

....

....[3]

[N09/2/5]

3. **Fig. (a)** shows a section of an alveolus.

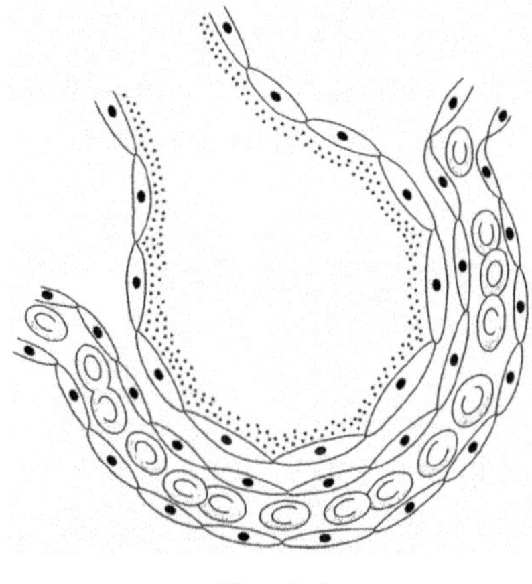

Fig. (a)

(a) Draw arrows on the diagram to show the movement of oxygen and carbon dioxide between the blood and the air in the alveolus. Label the arrows to show clearly which refers to oxygen and which refers to carbon dioxide. [2]

(b) **Fig. (b)** shows the oxygen uptake and lactic acid concentration before and during a period of exercise.

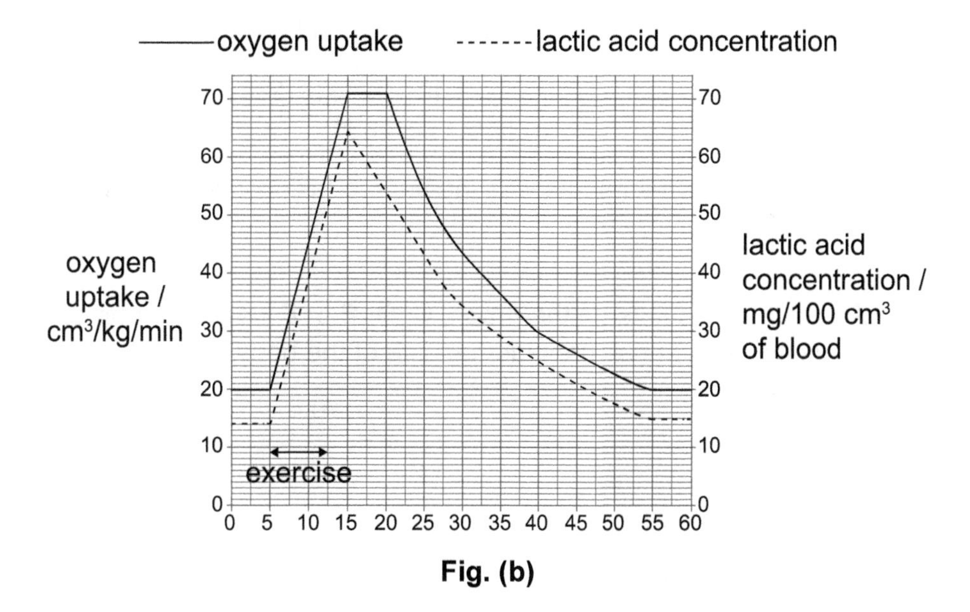

Fig. (b)

(i) State the difference between the maximum and minimum oxygen uptake.

....[1]

(ii) How long after the end of the exercise period did it take for the oxygen uptake to return to the resting level?

....[1]

(iii) Name the process which produces lactic acid.

....[1]

(iv) Explain why the oxygen uptake remains at a high level in the five minutes following the end of the period of exercise.

....

....[2]

[N08/2/2]

4. The figure shows the changes in the volume of the lungs of a person at rest over a period of 20 seconds.

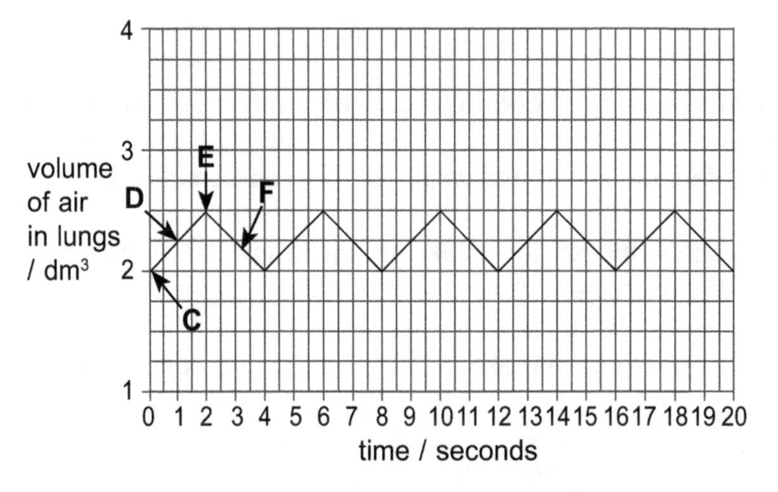

(a) How many breaths per minute was the person taking when at rest?

....[1]

(b) (i) At which **two** points on the graph, **C**, **D**, **E** or **F** was the air pressure in the lungs equal to the air pressure outside of the body?

....

....[1]

(ii) Explain your answer.

....

....[1]

(c) (i) Describe what is happening to the volume of air in the lungs between points **C** and **E**.

....

....[1]

(ii) Explain how this change is brought about.

....

....

....

....

....[4]

(d) A person runs up a hill for 12 seconds. On the graph in the figure, sketch a line to show the changes in the volume of air in the lung when running up the hill. [2]

[N06/2/4]

5. The figure shows the structures involved in oxygen uptake in the lungs.

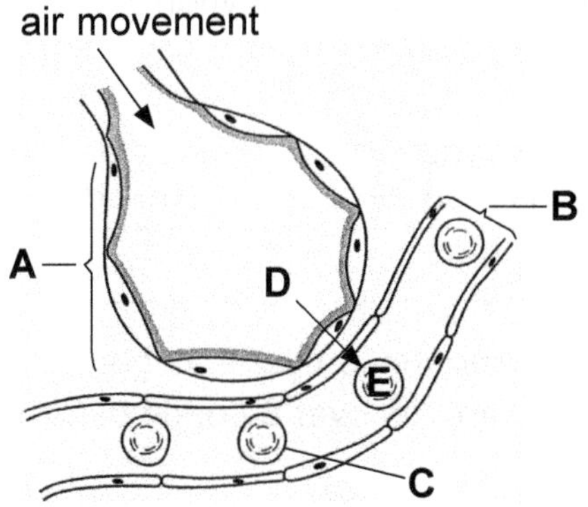

(a) Identify structures **A**, **B** and **C** in the figure.

A ....

....

B ....

....

C ....

....

[3]

(b) Each statement below describes a process that occurs during breathing.

Place a tick (✓) in the box beside each statement that describes a process necessary to cause air to move in the direction shown in the figure.

☐	diaphragm relaxes
☐	(external) intercostal muscles contract
☐	ribs rise
☐	diaphragm rises
☐	volume of thorax decreases

[2]

(c) Describe what happens to a molecule of oxygen as it moves from **D** to **E** in the figure.

....

....

....

....[3]

(d) The table shows the percentage of oxygen in the inspired air and expired air of a healthy person.

% oxygen in inspired air	% oxygen in expired air
20.5	16.5

Suggest and explain how these figures might be different for a person whose diet had been deficient in iron over a period of several years.

....

....

....

....[3]

[N05/2/2]

FREE RESPONSE QUESTIONS. . . . .

1. The oxygen content of blood in an artery and the oxygen used by a person were measured at rest and at increasing levels of exercise.

The figure shows the results.

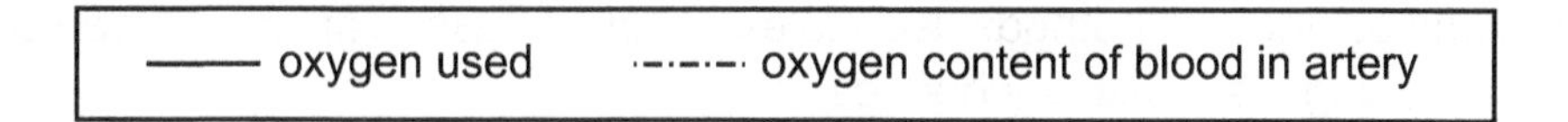

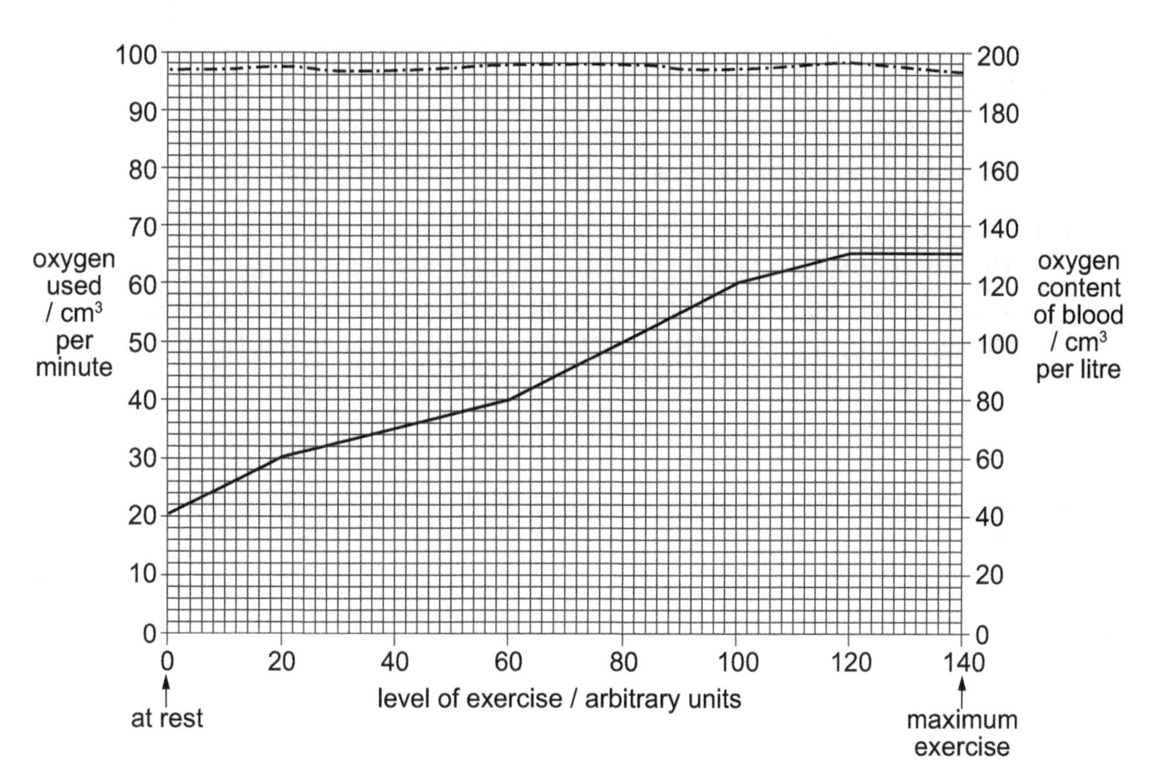

The oxygen content of blood in a vein was also measured and is shown in the table.

level of exercise / arbitrary units	0	20	40	60	80	100	120	140
oxygen content of blood / cm^3 per litre	164	120	96	76	64	52	44	40

(a) Plot the data from the table on the figure. [4]

(b) State the increase in the oxygen used from rest to maximum exercise. [1]

(c) State the difference between the oxygen content of blood in the artery and blood in the vein at maximum exercise. [1]

(d) Explain why the oxygen used increases as the level of exercise increases. [4]

[N12/2/8]

2. Carbon monoxide is found in cigarette smoke and when inhaled enters the blood and combines with haemoglobin.
Suggest how this information may be used to discourage pregnant mothers from smoking during pregnancy. [3]

[N08/2/10(c)]

3. Suggest and outline the methods used to establish the association between smoking tobacco and lung diseases, such as cancer. [3]

[N07/2/6(c)]

4. **(a) (i)** State two ways, other than muscle contraction, in which the body uses energy. [2]

(ii) How does anaerobic respiration differ from aerobic respiration? [2]

(b) Suggest explanations for the following:
'During a sprint race athletes get 80% of their energy from anaerobic respiration and 20% from aerobic respiration. During a long distance race athletes get 80% of their energy from aerobic respiration and 20% from anaerobic respiration.' [6]

[N07/2/7]

5. Explain how smoking may affect the health of the mother and the development of her fetus. [6]

[N06/2/8 Or (b)]

TOPIC 9

Excretion in Humans

MULTIPLE CHOICE QUESTIONS . . . .

1. In the kidney tubule, some substances are filtered and some substances are selectively reabsorbed.

Which row is correct?

	substance filtered	substance selectively reabsorbed
A	glucose	urea
B	protein	water
C	urea	glucose
D	water	protein

( )

[N12/1/20]

2. What happens when the brain secretes more anti-diuretic hormone (ADH)?

	water absorption by the kidney tubule	urine formation
A	less	less
B	less	more
C	more	less
D	more	more

( )

[N12/1/21]

3. The diagram shows a kidney and its associated vessels.

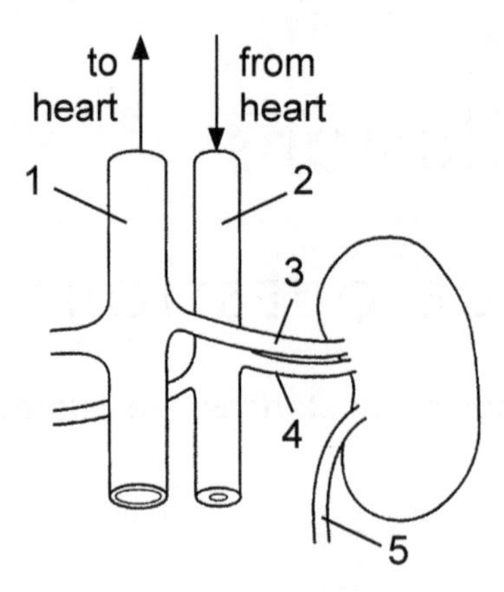

Which structures contain liquids with the **least** and the **most** concentrations of urea solution?

	least	most
A	1	4
B	2	1
C	3	4
D	3	5

( )

[N11/1/14]

4. What provides the force for ultra-filtration in the kidney tubule?

A anti-diuretic hormone (ADH)

B breakdown of urea

C contraction of the left ventricle

D mitochondria in the cells of Bowman's Capsule

( )

[N11/1/20]

5. The diagram represents the arm of a patient who is being treated by a machine.

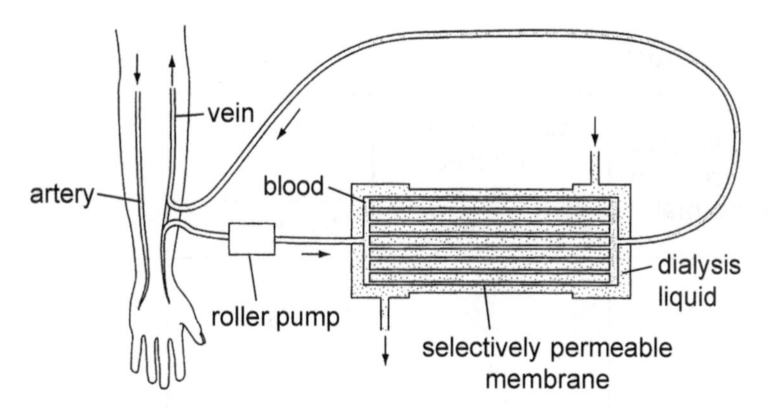

Which blood measurement does the machine affect?

A blood pressure
B carbon dioxide concentration
C oxygen concentration
D urea concentration

( )

[N11/1/21]

6. The diagram shows part of the urinary system of a mammal. Liquids pass through tubes **X**, **Y** and **Z** in the directions shown by the arrows.

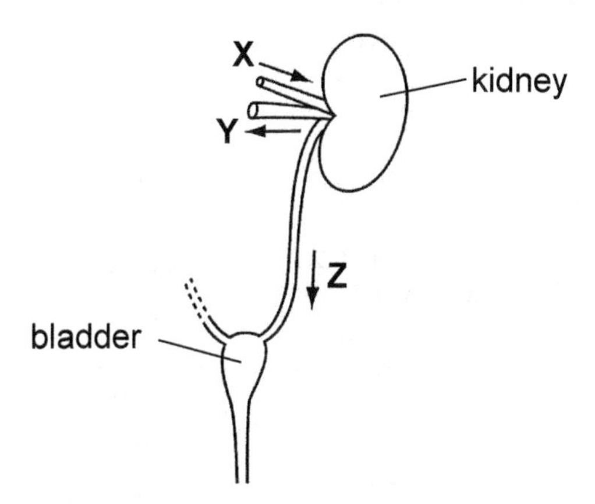

The volume of liquid passing through **Y** in one day is

A greater than that passing through **X**.
B less than that passing through **Z**.
C much less than that passing through **X** but slightly greater than that passing through **Z**.
D slightly less than that passing through **X** but much greater than that passing through **Z**.

( )

[N10/1/20]
[N07/1/20]

7. The table gives the events involved in the secretion and action of anti-diuretic hormone (ADH).

Which row is correct?

	water level in blood relative to normal	amount of ADH produced relative to normal	amount of water reabsorbed by kidneys
A	+	+	–
B	+	–	+
C	–	+	+
D	–	–	–

key
+ = increased
– = decreased

( )
[N10/1/21]

8. The most accurate description of excretion is that the body is eliminating

A the waste products of metabolism.
B undigested food materials from the intestines.
C unwanted products of respiration.
D water and urea from the kidneys.

( )
[N09/1/20]

9. What happens when a person drinks a large volume of water?

	amount of ADH secreted	re-absorption of water from kidney tubule	volume of urine produced
A	less	less	more
B	less	more	less
C	more	less	more
D	more	more	less

( )
[N09/1/21]

10. What is an example of excretion?

A release of insulin from the pancreas
B release of urea from the liver
C removal of carbon dioxide from the lungs
D removal of faeces from the alimentary canal ( )

[N08/1/20]

11. The diagram represents the process of dialysis in a kidney machine.

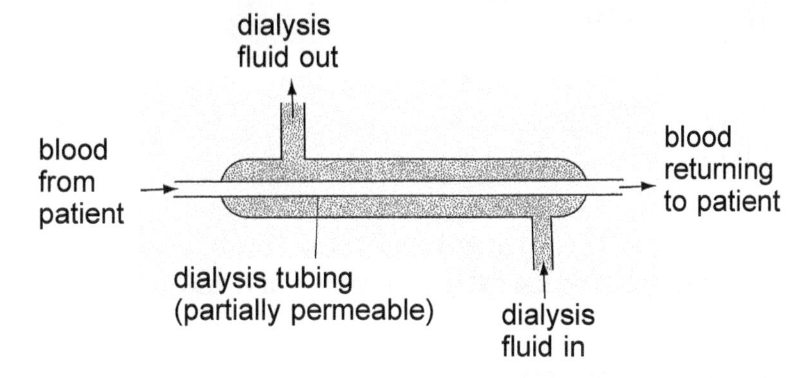

Which substance would **not** be present in the dialysis fluid flowing in?

A glucose
B mineral ions
C urea
D water ( )

[N08/1/21]

12. The diagram shows the flow of blood and dialysis fluid through a kidney machine.

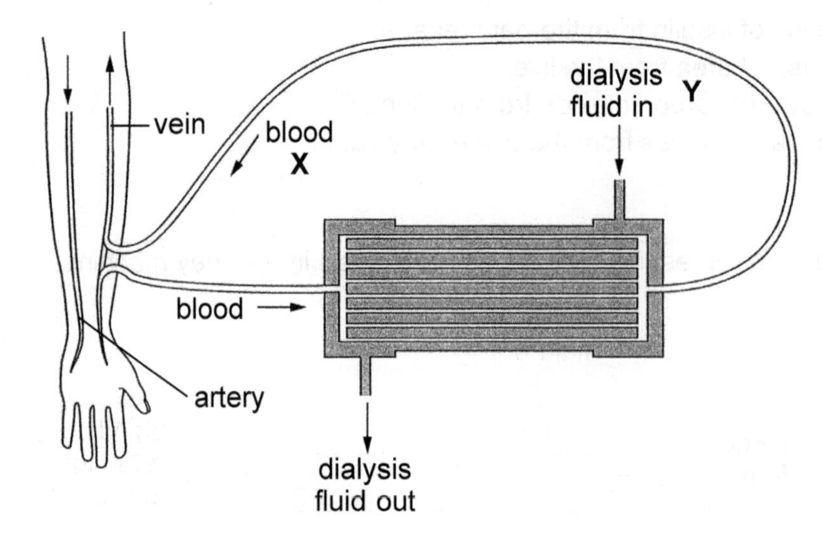

Which substances have the lowest concentration at **X** and the highest concentration at **Y**?

	lowest at **X**	highest at **Y**
A	glucose	salts
B	salts	glucose
C	urea	water
D	water	urea

( )

[N06/1/20]

13. What passes through the membranes of a kidney machine?

A protein and red blood cells
B urea and red blood cells
C water and protein
D water and urea

( )

[N05/1/22]

STRUCTURED QUESTIONS . . . .

1. For one week, three groups of 10 people were fed a diet containing measured amounts of protein.

On the last day of that week the amount of urea in their urine was measured. **Table (a)** shows the results of the analysis.

Table (a)

amount of urea excreted / g per day		
low protein diet	normal protein diet	high protein diet
4.75	19.20	31.50

(a) State why groups of 10 people rather than individuals were used in this investigation.

....[1]

(b) With reference to **Table (a)**, describe, using appropriate figures, what effect increasing the amount of protein in the diet had on the amount of urea in the urine.

....

....[2]

(c) **Table (b)** shows the relative composition of blood plasma and urine in humans.

Table (b)

substance	amount in plasma / g per 100 cm^3	amount in urine / g per 100 cm^3	concentration factor in urine
protein	8.500	0.000	-
urea	0.030	1.800	× 60.0
glucose	0.100	0.000	-
potassium	0.020	0.150	× 7.5
calcium	0.010	0.015	
chloride	0.330	0.600	× 1.8

(i) Explain the absence of protein and glucose in the urine.

protein..

....

glucose..

....

....[4]

(ii) Complete Table 2.2 for calcium. [1]

[N12/2/2]

2. **(a)** Define the term *excretion* and state its importance in the functioning of the body.

....

....

....

....[3]

(b) **Fig. (a)** shows part of the excretory system.

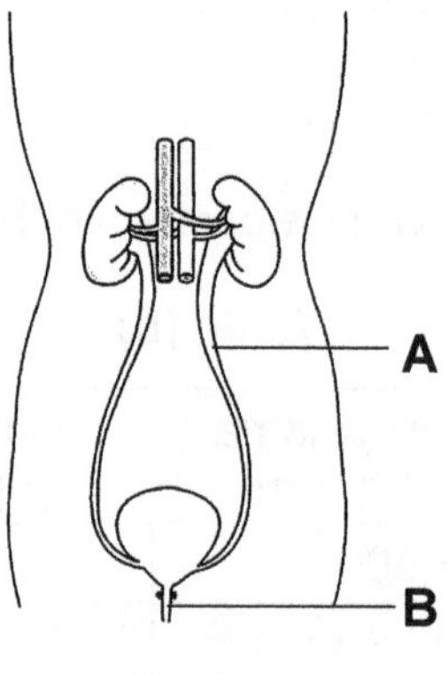

Fig. (a)

(i) Name the structures labelled **A** and **B**.

A ....

....

B ....

....

[2]

(ii) Name the nitrogen-containing excretory substance found in the liquid in **A**.

....[1]

(c) **Fig. (b)** shows how the pressure inside the bladder changes as it fills with urine.

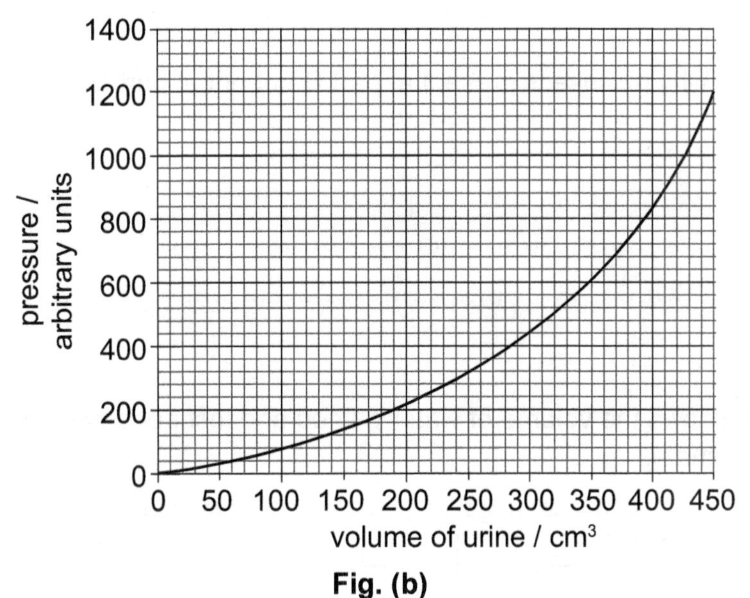

Fig. (b)

(i) State the pressure when the bladder contains 400 cm^3 of urine.

....[1]

(ii) When the bladder contained 450 cm^3 of urine the person urinated 350 cm^3 of urine.
State by how much the pressure in the bladder changed.

....[1]

[N11/2/4]

3. The figure shows a kidney tubule.

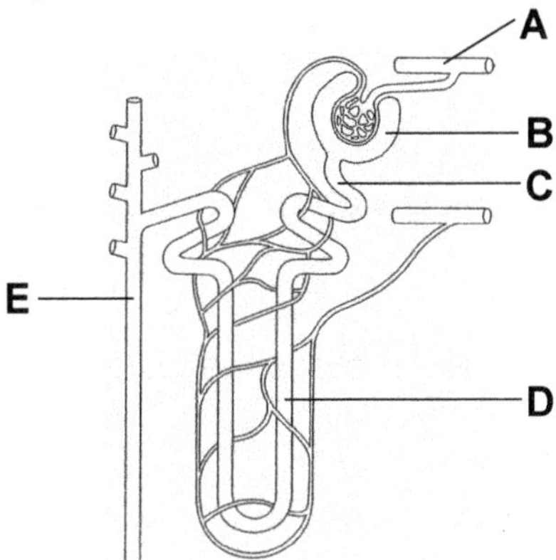

(a) Name two components of the blood found in **A** but not in **B**.

1. ....

....

2. ....

....[2]

(b) Name **one** substance, other than water, present at parts **A**, **B**, **C**, **D** and **E**.

....[1]

(c) Explain why glucose is present at part **B** but not at part **D**.

....

....

....

....[3]

[N08/2/5]

4. The figure shows the volume of water gained and lost per day by a person living in a tropical climate.

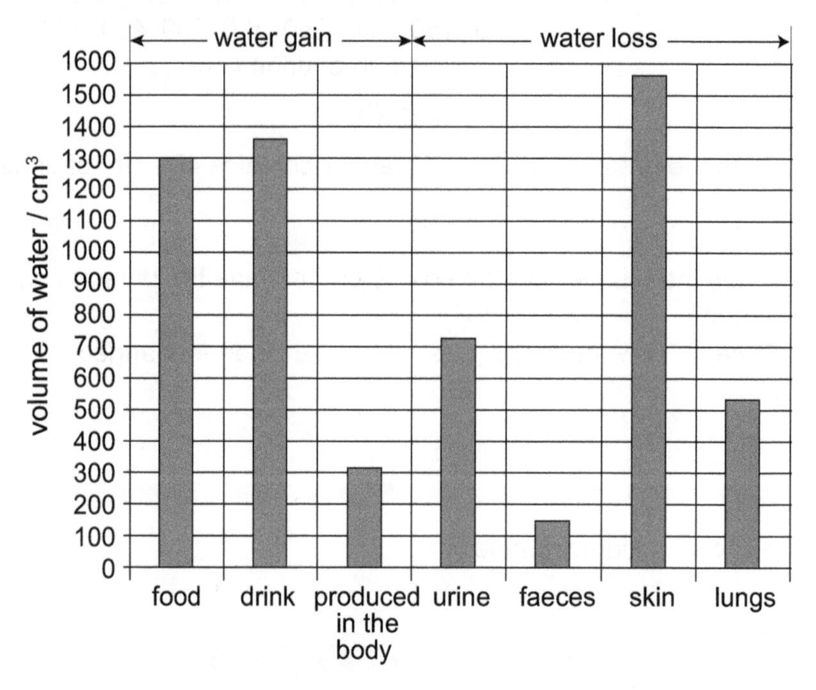

(a) **(i)** State the form in which water is lost from the lungs.

....[1]

(ii) Name the process which produces water in the body.

....

....[1]

(b) (i) With reference to the figure, state the total volume of water gained in a day.

(You may use the space below for your working.)

....[1]

(ii) Calculate the total volume of water lost per day and the percentage of this which is lost in faeces.

(You may use the space below for your working.)

....[1]

(c) Suggest and explain two ways in which the figures in the bar chart would change in a cold climate.

....

....

....

....[4]

(d) Approximately 190 dm^3 of water is filtered through the kidneys each day.

Use this information and the figures in the bar chart to explain what happens to this water in the kidney.

....

....

....

....[2]

[N07/2/3]

FREE RESPONSE QUESTIONS. . . . .

1. Describe and discuss the process of dialysis using a kidney machine. [7]

[N06/2/7(b)]

TOPIC 10
Homeostasis

MULTIPLE CHOICE QUESTIONS . . . .

1. The diagram shows a section through the skin.

On a cold day, which labelled part of the skin will be warmest?

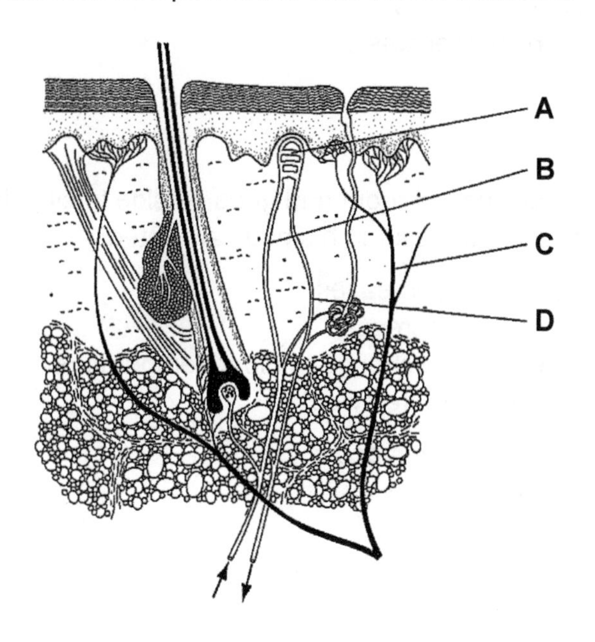

( )

[N12/1/22]

2. The diagram shows an example of homeostasis in a person.

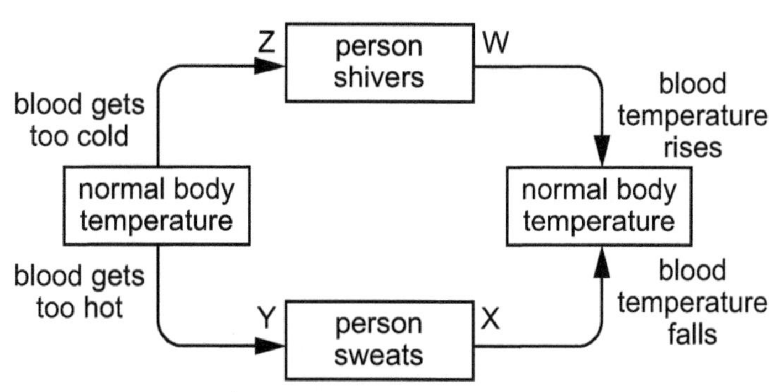

Which two letters represent negative feedback changes?

A W and X
B W and Y
C X and Z
D Y and Z

( )

[N12/1/23]

3. Which factors are controlled by homeostasis?

	glucose concentration in blood	water content in the ileum	temperature in the stomach	pH in the duodenum
A	✓	×	✓	×
B	✓	×	✓	✓
C	✓	✓	×	✓
D	×	✓	✓	×

key
✓ = controlled by homeostasis
× = not controlled by homeostasis

( )

[N11/1/22]

4. A man leaves a cool, shaded room and goes outside to sit in the hot sun. Some of the changes that occur in his body are listed.

1 Blood temperature decreases.
2 Blood temperature increases.
3 The brain detects the change in blood temperature.
4 The skin produces more sweat.

In which order do these events occur?

A 2 → 3 → 4 → 1
B 2 → 4 → 1 → 2
C 3 → 1 → 4 → 2
D 4 → 1 → 3 → 2

( )

[N11/1/23]

5. The diagram shows a section through the skin.

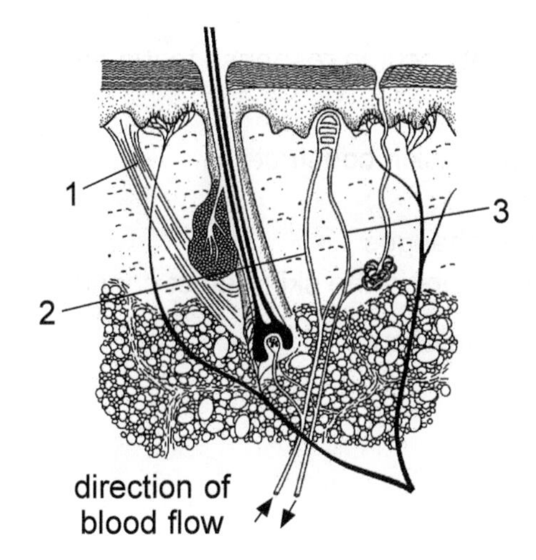

Which structure(s) contain muscle that contracts when the body is too cold?

A 1 only
B 1 and 2
C 1 and 3
D 2 only

( )

[N10/1/22]

6. Four processes that take place in the human body are listed.

1 absorption of amino acids through the villi
2 maintenance of constant body temperature
3 production of lactic acid in muscles
4 regulation of blood glucose concentration

Which two processes are directly controlled by negative feedback?

A 1 and 3
B 1 and 4
C 2 and 3
D 2 and 4

( )

[N10/1/23]

7. During which process does the body lose heat by evaporation of water?

A defecation
B sweating
C urination
D vomiting

( )

[N09/1/22]

8. A shop worker enters a walk-in freezer to carry out a stock check that takes about ten minutes.

Some of the changes that occur in response to the drop in external temperature are listed.

1 blood vessels in the skin constrict
2 brain takes action
3 skin temperature changes
4 temperature receptors in the skin detect change

Which sequence of events occurs?

A $2 \rightarrow 1 \rightarrow 3 \rightarrow 4$
B $3 \rightarrow 1 \rightarrow 4 \rightarrow 2$
C $4 \rightarrow 2 \rightarrow 1 \rightarrow 3$
D $4 \rightarrow 1 \rightarrow 2 \rightarrow 3$

( )

[N09/1/23]

9. Which factors are controlled by homeostasis?

	glucose concentration in blood	water content in the ileum	temperature in the stomach	pH in the duodenum
A	✓	✓	×	✓
B	✓	×	✓	✓
C	✓	×	✓	×
D	×	✓	✓	×

key
✓ = controlled by homeostasis
× = not controlled by homeostasis

( )

[N08/1/22]

10. The diagram shows a section through the skin.

Which structure detects changes in skin temperature?

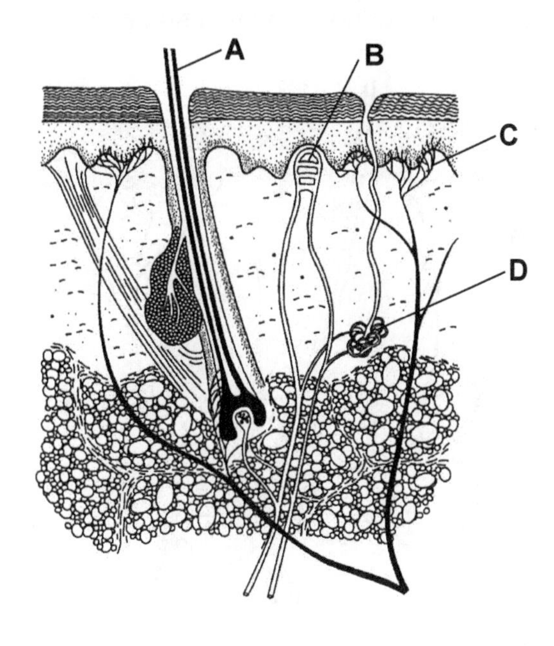

( )

[N07/1/21]

11. The body can regulate both its temperature and the amount of water in its cells.

What are these processes?

A assimilation
B excretion
C homeostasis
D osmosis

( )

[N06/1/21]

12. When the external temperature drops, the following changes may take place in the human body.

1 body temperature falls
2 body temperature rises
3 brain detects cooler blood
4 shivering begins

In which order do they occur?

	first	⟶		last
A	1	3	4	2
B	1	4	3	2
C	3	2	4	1
D	3	4	2	1

( )

[N06/1/22]

13. A person walks into a very cold room. Shortly afterwards the hairs on their skin are raised.

Which labelled structure is included in the first stage of this reflex?

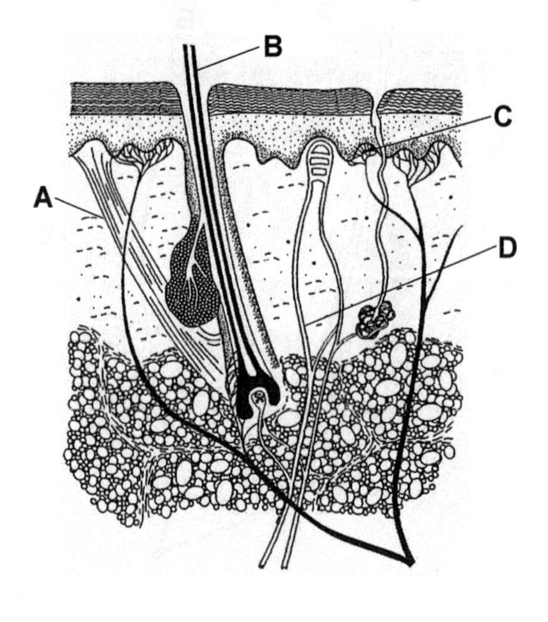

( )

[N05/1/19]

14. The circles represent the diameter of the blood vessels in the surface of the skin as the body temperature changes.

Which shows the diameter of the blood vessels after a decrease and after an increase in body temperature?

	diameter of blood vessels	
	after a decrease in body temperature	after an increase in body temperature
A		
B		
C		
D		

( )

[N05/1/21]

STRUCTURED QUESTIONS . . . .

1. **Fig. (a)** shows a vertical section of the skin.

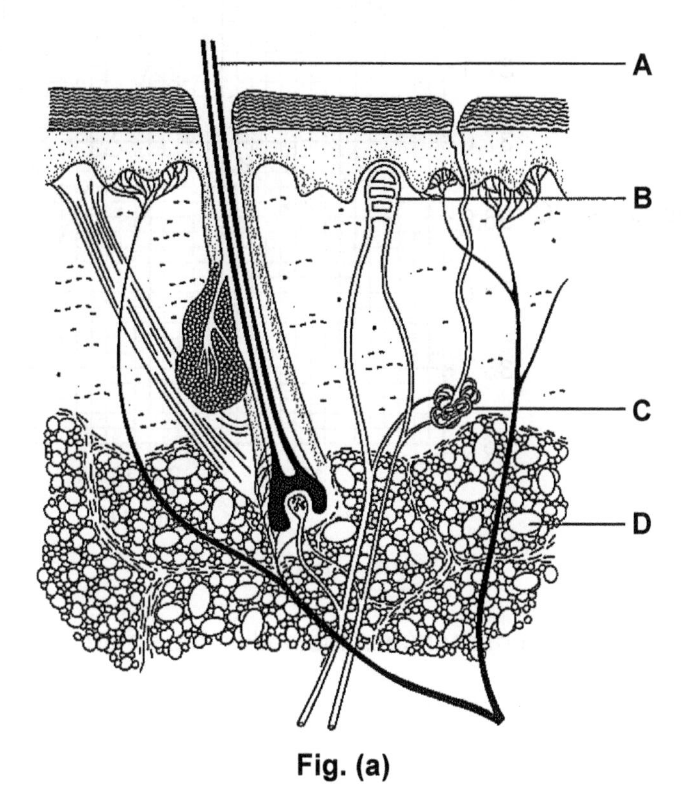

Fig. (a)

(a) **(i)** Identify the parts labelled **A**, **B** and **D**.

A...

B...

D...[3]

(ii) Describe the function of part **C** in temperature regulation.

....

....

....

....

....

....

....[3]

(b) **Fig. (b)** is a graph showing the changes in body temperature of a healthy person over a 24-hour period.

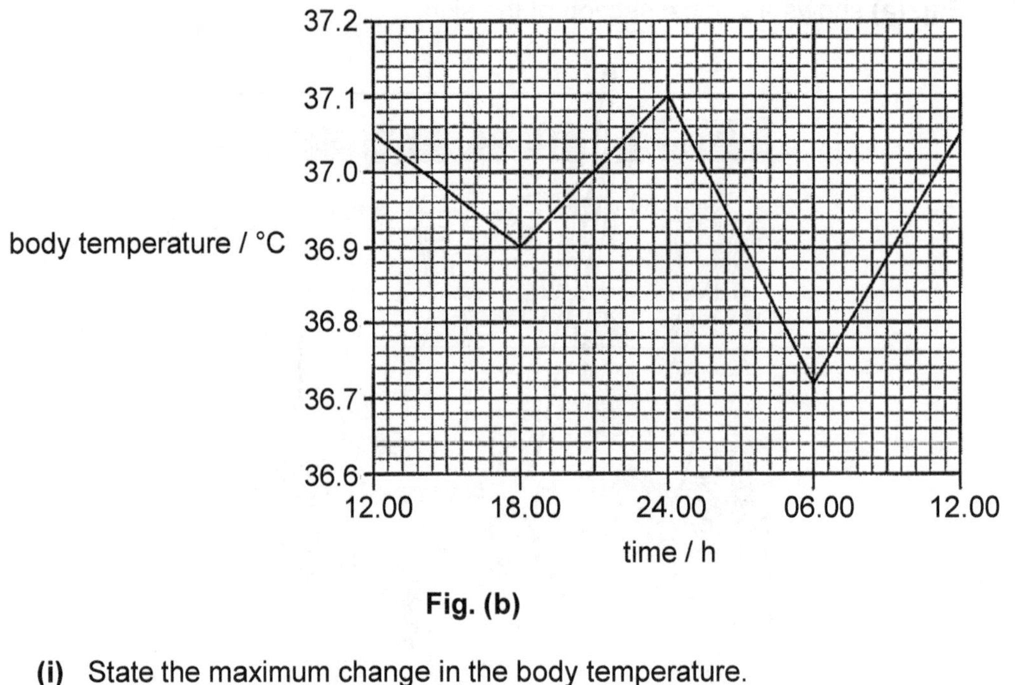

Fig. (b)

(i) State the maximum change in the body temperature.

....[1]

(ii) Suggest an explanation for the fall in body temperature between 24.00 and 06.00.

....

....[2]

[N12/2/3]

2. The table shows the average body length and rate of heat production in five animals.

animal	average body length / cm	rate of heat production / kJ kg^{-1} h^{-1}
rat	10	17.0
rabbit	30	13.0
dog	70	7.2
human	140	3.3
camel	180	3.0

(a) (i) Draw a graph to show the relationship between the rate of heat production and the average body length.

[4]

(ii) Describe the relationship between average body length and rate of heat production.

....[1]

(iii) Name the process that releases energy in the body.

....[1]

(b) (i) Some humans of average body mass of 74 kg stayed in a hot desert environment for six hours.

The average percentage of body mass lost per hour was 1.4%.

Calculate the total average body mass lost, in kg, over the six-hour period.

....[1]

(ii) Most of the body mass lost is in the form of water.

Explain how this loss of body mass helps survival in the hot desert environment.

....

....[3]

[N06/2/3]

FREE RESPONSE QUESTIONS. . . . .

1. **(a)** In extremely cold conditions people may get frostbite. This causes the cells in the toes and fingers to die.

Explain why this takes place even if thick gloves, socks and shoes are worn. [5]

(b) Cold environments cause the body to shiver.
Explain why shivering takes place in cold conditions. [3]

(c) Explain how the mechanisms for controlling body temperature are co-ordinated. [2]

[N08/2/11 Either]

2. Explain what is meant by the terms *homeostasis* and *negative feedback*.

(a) *homeostasis* [2]

(b) *negative feedback* [2]

[N08/2/11 Or (a)]

3. Define the term homeostasis.
From your knowledge of diabetes mellitus suggest how the gland that produces insulin may be considered as a homeostatic organ. [3]
[N06/2/7(a)]

TOPIC 11

Co-ordination and Response in Humans

MULTIPLE CHOICE QUESTIONS . . . .

1. When a person steps onto a drawing pin, the leg muscles contract to pull the foot away and pain is felt.

Which part of the nervous system co-ordinates the leg movement and which part perceives the pain?

	co-ordinator of leg movement	pain perceived by
A	spinal cord	brain
B	motor neurone	leg muscles
C	brain	sensory neurone
D	sensory neurone	spinal cord

( )

[N12/1/24]

2. The diagram shows what can happen to glucose in the human body.

Which change is controlled by glucagon?

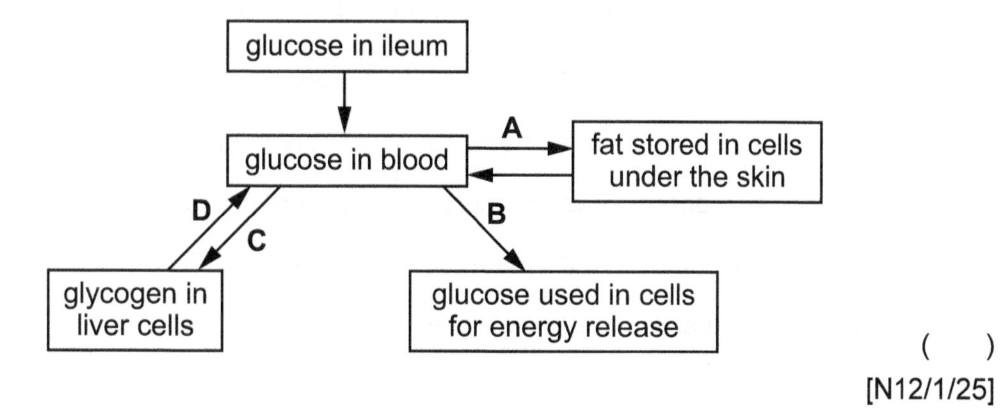

( )

[N12/1/25]

3. The graph shows changes in a student's blood glucose concentration over a four hour period.

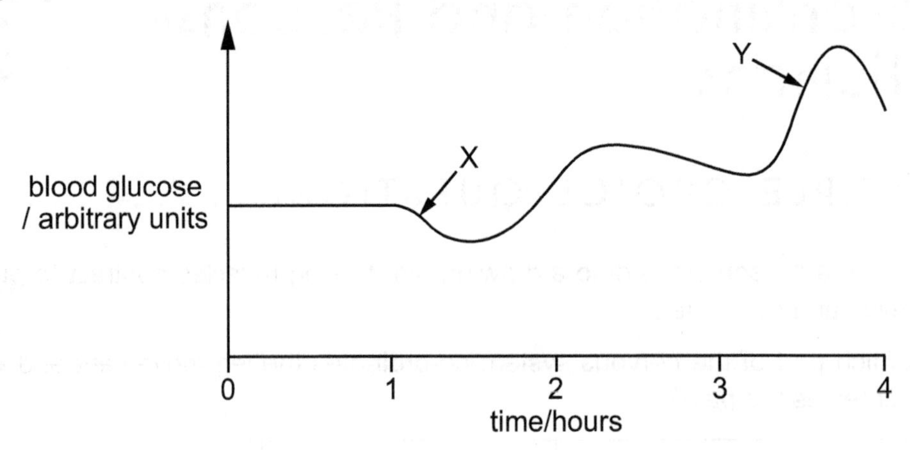

What causes the changes at X and Y?

	X	Y
A	decreased insulin	decreased adrenaline
B	decreased insulin	increased adrenaline
C	increased adrenaline	increased insulin
D	increased insulin	increased adrenaline

( )

[N12/1/26]

4. A child is frightened by a loud noise and shouts for help.

In which order are the different types of neurone involved in this response?

	involved first	⟶	involved last
A	motor neurone	relay neurone	sensory neurone
B	motor neurone	sensory neurone	relay neurone
C	sensory neurone	motor neurone	relay neurone
D	sensory neurone	relay neurone	motor neurone

( )

[N11/1/24]

5. A student is watching a football match. The diagram shows a **vertical** section through one of his eyes.

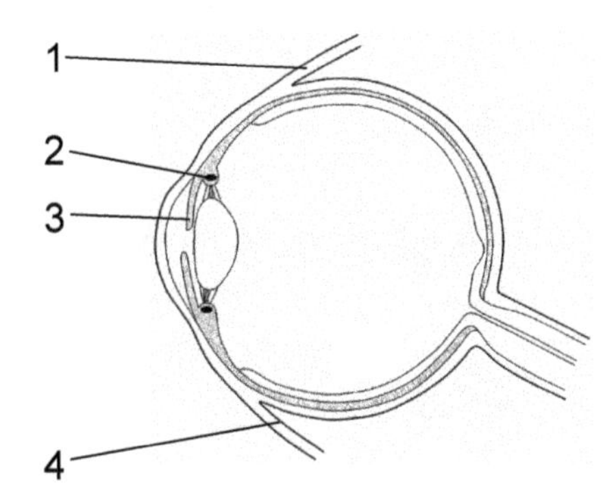

He then looks down at the wrist-watch on his arm to check if it is nearly half-time.

Which eye muscles contract?

A 1 and 2
B 1 and 3
C 2 and 4
D 3 and 4

( )

[N11/1/25]

6. In an experiment, a student is threatened by a large dog and has to run away to escape.

During the incident the adrenaline levels in the student's blood are measured. The graph shows the results.

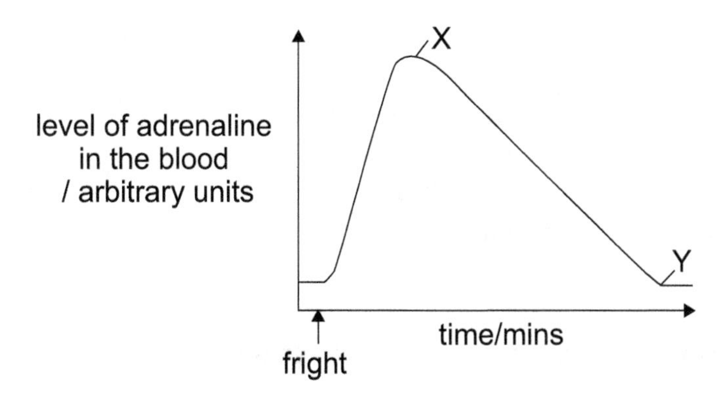

Which statement explains the change in adrenaline levels between points **X** and **Y**?

A Adrenaline is being broken down by the liver.
B Adrenaline is being excreted by the kidneys.
C Adrenaline is being returned to the endocrine gland that produced it.
D Adrenaline is being used up by the contracting muscles.

( )

[N11/1/26]

7. The diagram shows a section through the eye.

Which structure is mainly responsible for movement of the eyeball so that light falls on the fovea?

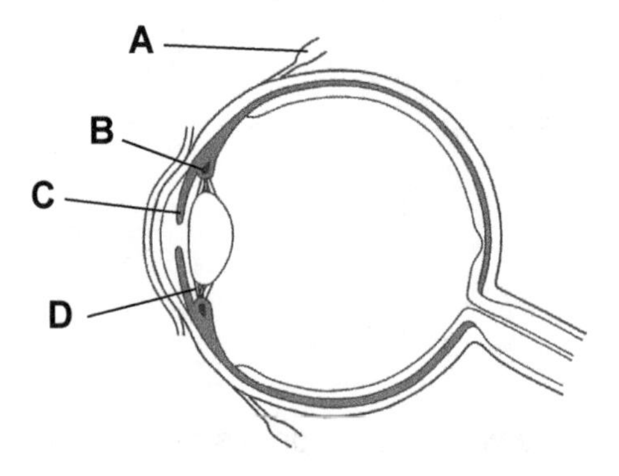

( )

[N10/1/24]

8. What effects would an increase in adrenaline have on the body?

	concentration of glycogen in the liver	concentration of glucose in the blood
A	decrease	increase
B	increase	increase
C	increase	no effect
D	no effect	decrease

( )

[N10/1/25]

9. Which information about insulin or glucagon is correct?

A Excess insulin results in diabetes mellitus.
B Glucagon is broken down in the liver.
C Glucagon is manufactured in the liver.
D Insulin raises the blood glucose level.

( )

[N10/1/26]

10. What co-ordinates a reflex action that requires a response from several parts of the body?

A brain
B endocrine glands
C muscles
D spinal cord

( )

[N09/1/24]

11. The diagram shows a section through the eye.

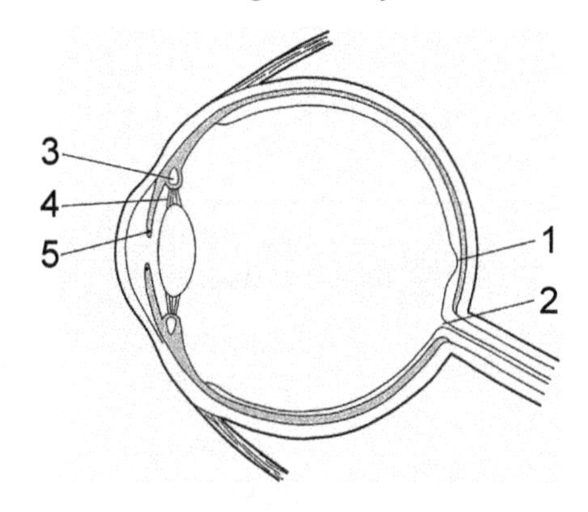

Which labelled structures are effectors and which are the receptors?

	effectors	receptors
A	1	4
B	3	2
C	4	3
D	5	1

( )

[N09/1/25]

12. The table gives information about endocrine glands.

Which information is correct?

	gland	hormone produced	target organ	effect
A	adrenal	adrenaline	liver	decreases blood glucose level
B	ovaries	progesterone	uterus	ovulation occurs
C	pancreas	insulin	liver	conversion of glucose to glycogen
D	testes	testosterone	penis	becomes erect to allow sexual intercourse

( )

[N09/1/26]

13. How is the concentration of glucose in the blood regulated?

	blood glucose concentration	pancreas stimulated to secrete	liver converts	blood glucose concentration	pancreas reduces secretion of
A	fall	glucagon	glycogen to glucose	rise	glucagon
B	fall	insulin	glucagon to glucose	rise	insulin
C	rise	glucagon	glucose to glycogen	fall	glucagon
D	rise	insulin	glycogen to glucose	fall	insulin

( )

[N08/1/23]

14. The diagrams show two sections through the eye of the same person.

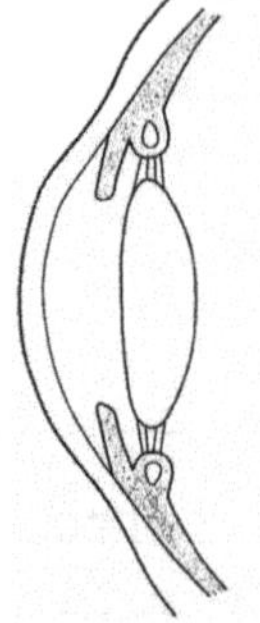

diagram 1
focusing on an object
sixty metres away
in daylight

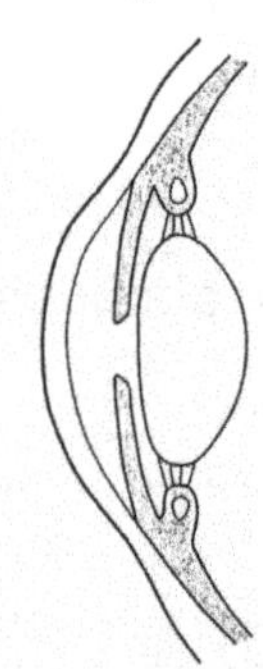

diagram 2
focusing on an object
one metre away
in very bright light

What happens to achieve the changes from the eye in diagram 1 to the eye in diagram 2 under the different conditions?

	ciliary muscles	iris radial muscles	iris circular muscles
A	contract	contract	relax
B	contract	relax	contract
C	relax	contract	relax
D	relax	relax	contract

( )

[N08/1/24]

15. The diagram shows a section of the spinal cord and spinal nerve roots.

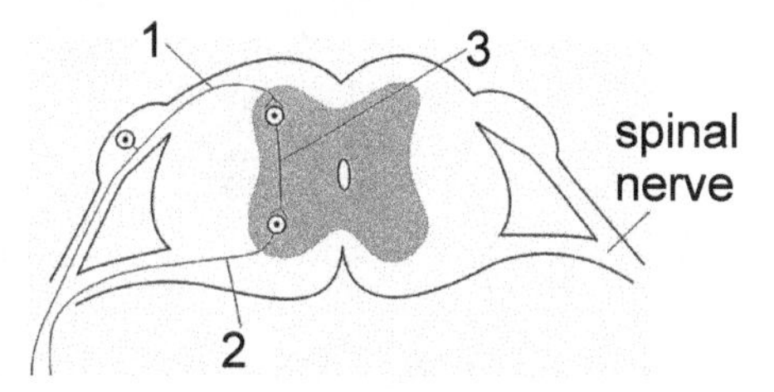

What identifies the neurones of the reflex arc shown?

	motor neurone	relay neurone	sensory neurone
A	1	2	3
B	1	3	2
C	2	1	3
D	2	3	1

( )

[N08/1/25]

16. Hormones are chemicals involved in co-ordination in the body.

Which combination in the table is correct?

	hormones are carried by	hormones are destroyed by
A	blood plasma	kidney
B	red blood cells	kidney
C	blood plasma	liver
D	red blood cells	liver

( )

[N08/1/26]

17. The flow diagram shows the pupil reflex.

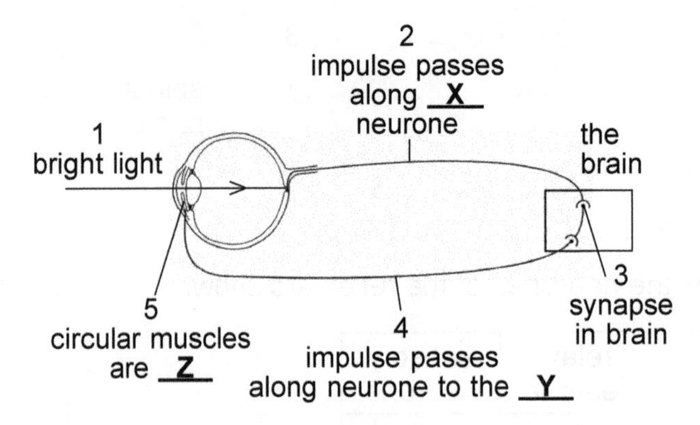

Which words complete the flow diagram?

	X	Y	Z
A	motor	ciliary body	contracting
B	motor	iris	relaxing
C	sensory	ciliary body	relaxing
D	sensory	iris	contracting

( )
[N07/1/22]

18. A railway traveller sees his train approaching from the distance, looks at his watch and checks it with the station clock.

Which of the following shows the sequence of changes taking place in the shape of the lens of his eye?

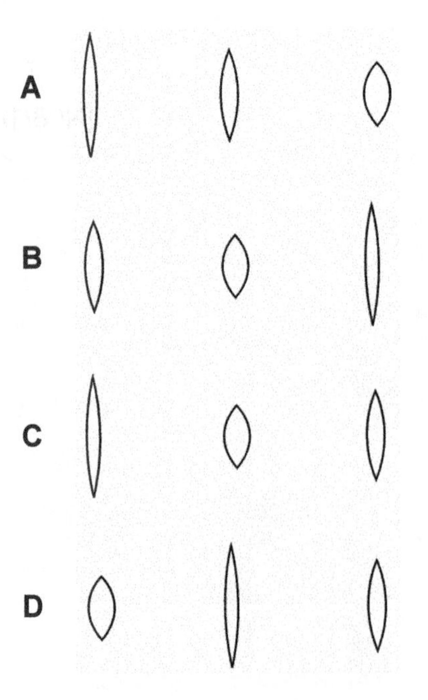

( )
[N07/1/23]

19. Three directions in which nerve impulses can travel in the nervous system are listed.

1 away from the central nervous system
2 towards the central nervous system
3 within the central nervous system

In which direction do impulses in sensory and relay (intermediate) neurones travel?

	sensory neurone	relay neurone
A	1	2
B	1	3
C	2	1
D	2	3

( )
[N06/1/23]

20. What structures cover the pupil of a human eye?

A conjunctiva and cornea
B conjunctiva and sclera
C cornea and retina
D retina and sclera

( )
[N06/1/24]

21. The list shows events when a person goes from a dark room out into bright sunlight.

1 circular iris muscles contract
2 increase in impulses in optic nerve to brain
3 more light strikes the retina of the eye
4 pupil diameter decreases

In which order do these events occur?

	first	⟶		last
A	1	4	2	3
B	1	2	3	4
C	3	2	1	4
D	3	1	4	2

( )
[N06/1/25]

22. The diagram shows a section of a human eye focused on a near object.

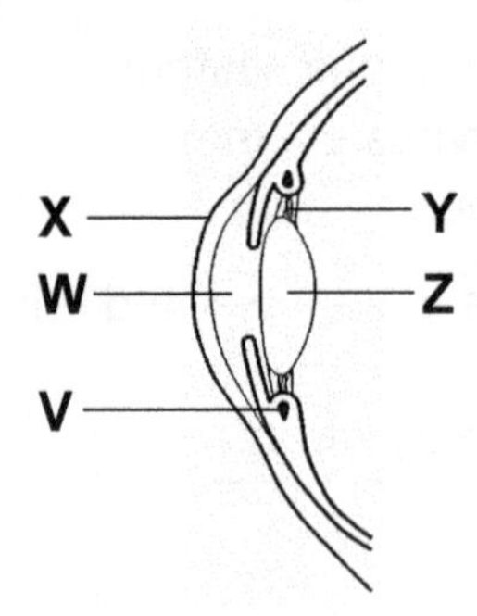

Which parts will change to focus on a distant object?

A	**W**	**X**	**Y**
B	**V**	**X**	**Z**
C	**W**	**Y**	**Z**
D	**V**	**Y**	**Z**

( )

[N05/1/23]

23. The graphs show the concentrations of glucose and insulin in the blood of a healthy person.

Which graph shows the changes expected after a meal containing starch?

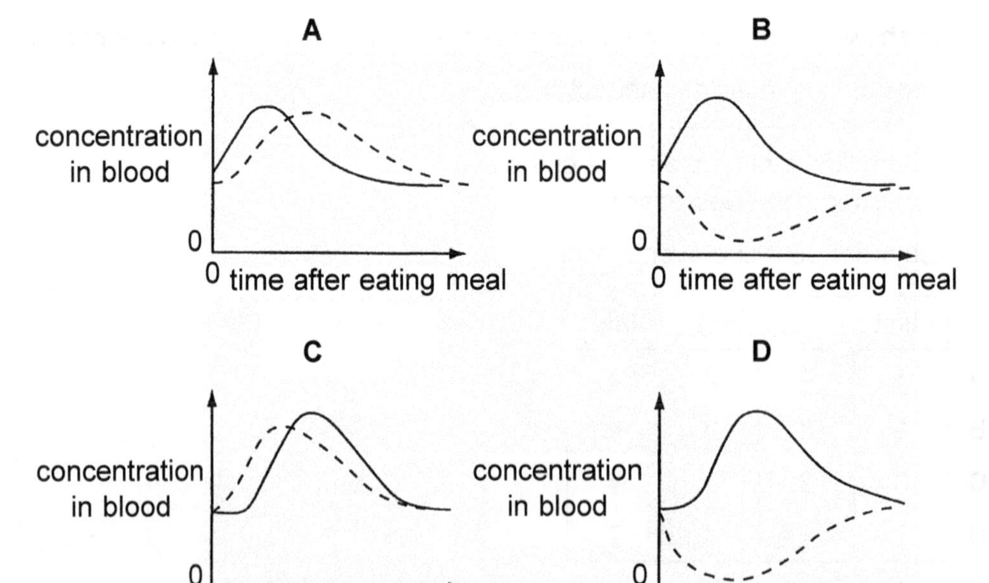

( )

[N05/1/26]

STRUCTURED QUESTIONS . . . .

1. The figure shows some of the features of how the body is co-ordinated by the nervous and hormonal systems.

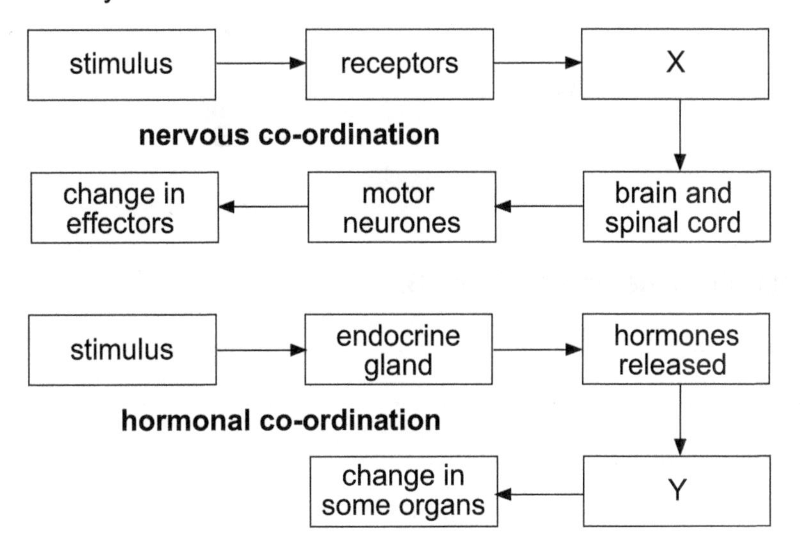

(a) **(i)** Name **X** and **Y**.

X..

....

Y..

....[2]

(ii) State **one** way in which nervous and hormonal co-ordination are similar.

....

....[1]

(iii) State **one** way in which nervous and hormonal co-ordination are different.

....

....[1]

(b) **(i)** Name the stimulus which would cause the release of insulin.

....

....[1]

(ii) Name the effectors that would respond to the stimulus of bright light shining into the eye.

....

....[1]

[N10/2/6]

2. The figure shows a horizontal section of the eye.

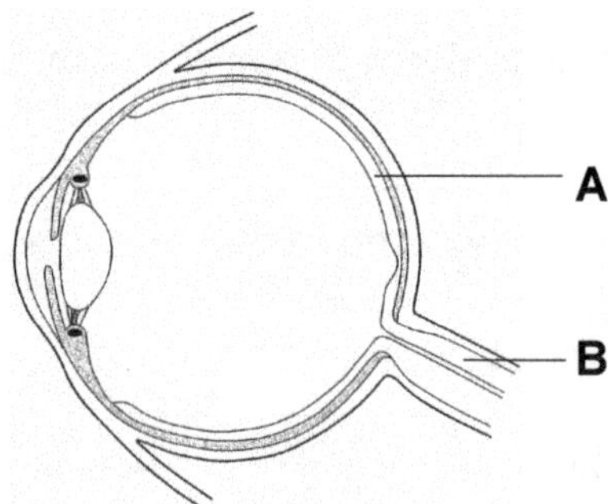

(a) Identify the parts labelled **A** and **B**.

A..

....

B..

....
[2]

(b) A student wrote 'messages pass along structure **B**'.

(i) State the correct term for these 'messages'.

....

....[1]

(ii) State the organ to which these 'messages' are carried.

....

....[1]

(c) The table contains statements which describe some of the changes in the eye when a person looks closely at a book after looking at a distant object.

Using the numbers 1 – 4, indicate the correct sequence of these changes starting from when the eye is focused on the distant object.

lens is thick	
suspensory ligaments slacken	
lens is thin	
ciliary muscles contract	

[2]

(d) The lens of the eye is naturally transparent. Cataracts are a gradual, painless clouding of this clear lens.

Suggest how cataracts may affect vision.

....

....

....

....

....

....[4]

[N06/2/1]

FREE RESPONSE QUESTIONS. . . . .

1. **(a)** Describe the similarities and differences between a voluntary action and a reflex action. [4]

   (b) Describe the pathway of nerve impulses in a **named** reflex action. [6]

   [N12/2/10 Either]

2. The figure shows a section of the eye.

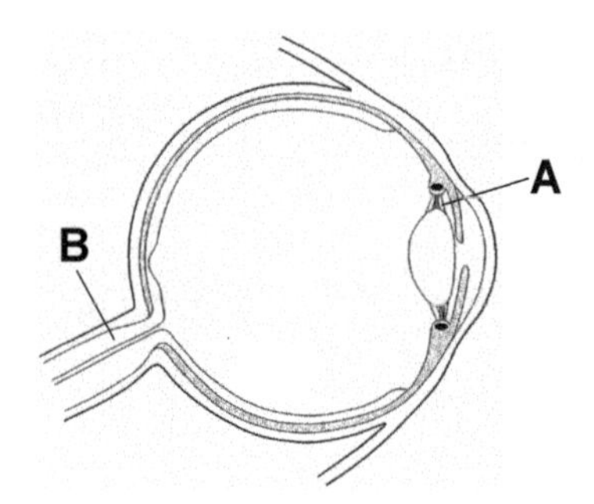

   (a) Name and state the function of the parts labelled **A** and **B**.

   (i) **A** name and function [2]

   (ii) **B** name and function [2]

   (b) Describe how the parts of the eye produce a focused image of a distant object. [5]

   [N09/2/9]

3. Explain how the concentration of glucose in the blood is kept within narrow limits. [6]

   [N08/2/11 Or (b)]

4. **(a)** **(i)** Define a hormone. [3]

   (ii) Give an example of a situation in which there would be a rapid rise in the concentration of adrenaline in the blood.

   What would be the effect of this rise? [2]

   (b) Describe the similarities and differences between a voluntary action and a reflex action. [5]

   [N07/2/8 Or]

5. A person is sitting in the shade reading a book when he looks at the bright sky to see an aeroplane flying past. Explain the changes in

   (a) the lens, and [6]

   (b) the pupil. [4]

   [N05/2/6]

TOPIC 12
Reproduction

MULTIPLE CHOICE QUESTIONS . . . .

1. Which statement about reproduction in plants is correct?

A All the adult offspring formed by asexual reproduction have identical phenotypes.
B Asexual reproduction only occurs in plants.
C Offspring formed after cross-pollination contain genetic material from two individuals.
D Sexual reproduction always produces gametes by the fusion of zygotes.

( )
[N12/1/27]

2. Flowering plants use different methods to ensure that their flowers are pollinated successfully.

Some of these methods are listed.

1 Plant 1 has flowers in which the female parts ripen before the male parts.
2 Plant 2 has separate male and female flowers.
3 Plant 3 has separate male and female plants.
4 Plant 4 has flowers in which the male parts ripen before the female parts.
5 Plant 5 has flowers in which the male and female parts ripen at the same time.

Which method(s) make it more likely that cross-pollination will take place?

A 1, 2, 3 and 4 only
B 1 and 4 only
C 2 and 3 only
D 5 only

( )
[N12/1/28]

3. Which row shows the effects of estrogen and progesterone?

	high levels needed for ovulation	high levels needed to stop development of more ova	maintains the uterus lining	repairs the uterus lining
A	estrogen	progesterone	estrogen	progesterone
B	estrogen	progesterone	progesterone	estrogen
C	progesterone	estrogen	estrogen	progesterone
D	progesterone	estrogen	progesterone	estrogen

( )
[N12/1/29]

4. Some constituents of the blood are listed.

1 amino acids
2 antibodies
3 carbon dioxide
4 glucose
5 minerals
6 oxygen
7 red blood cells
8 urea

Which constituents pass from baby to mother across the placenta and which pass from mother to baby?

	baby to mother	mother to baby
A	2 and 3	1 and 8
B	3 and 4	5 and 7
C	3 and 8	1 and 2
D	5 and 6	4 and 3

( )

[N12/1/27]

5. Which characteristics does a wind-pollinated flower have?

A anthers with short filaments
B nectaries producing nectar
C short, unbranched stigmas
D small, smooth pollen grains ( )

[N11/1/27]

6. After fertilisation, which structure develops into the seed of a flowering plant?

A carpel
B ovule
C ovum
D receptacle ( )

[N11/1/28]

7. What is an advantage of the testes being held in the scrotum, outside the body cavity?

A More sperm can be stored in an external scrotum.
B Sperm development is more efficient at temperatures below 36°C.
C Testes are better protected in the scrotum than in the body cavity.
D There is more time for prostate secretions to be added to sperm. ( )

[N11/1/29]

8. Which statement about the methods of reproduction is correct?

A All the offspring from sexual reproduction are genetically identical but dissimilar to the parents.

B Asexual reproduction results in identical zygotes.

C In asexual reproduction, two nuclei from one parent fuse.

D In sexual reproduction, one nucleus from each of the two parents fuse.

( )

[N11/1/30]

9. What would be found in an insect-pollinated flower?

A a feathery stigma

B long hanging stamens

C petals reduced or absent

D sticky pollen grains

( )

[N10/1/27]

10. The diagram shows a section through a flower.

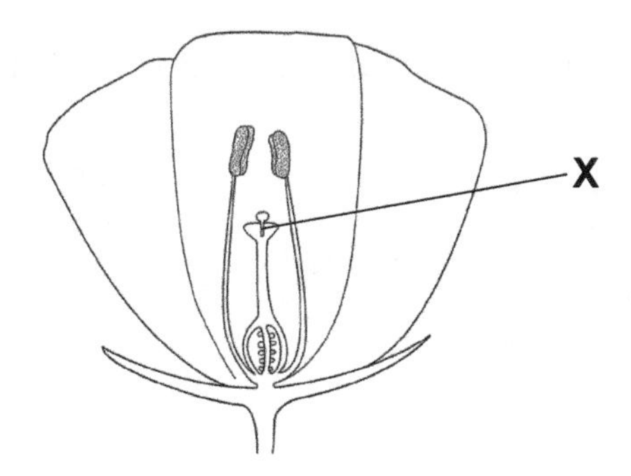

What passes down tube **X**?

A female gamete

B male gamete

C nectar

D pollen grain

( )

[N10/1/28]

11. The diagram shows a section through the male reproductive system.

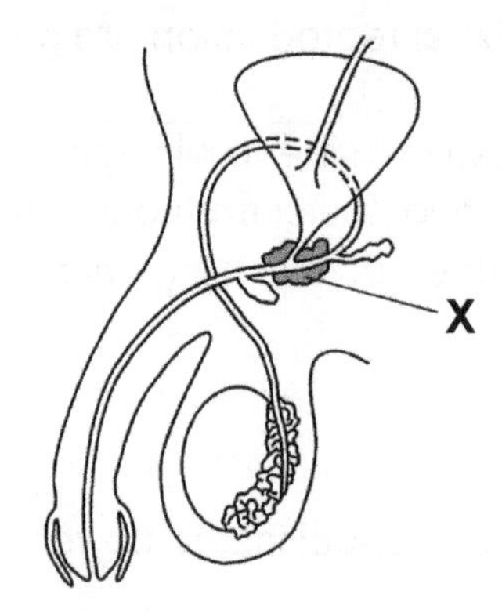

What will be the effect if gland **X** is removed?

A Fewer sperm are formed.
B Fewer sperm can be stored.
C Less testosterone is produced.
D Sperm are less active.

( )
[N10/1/29]

12. The diagram shows a section through part of the placenta.

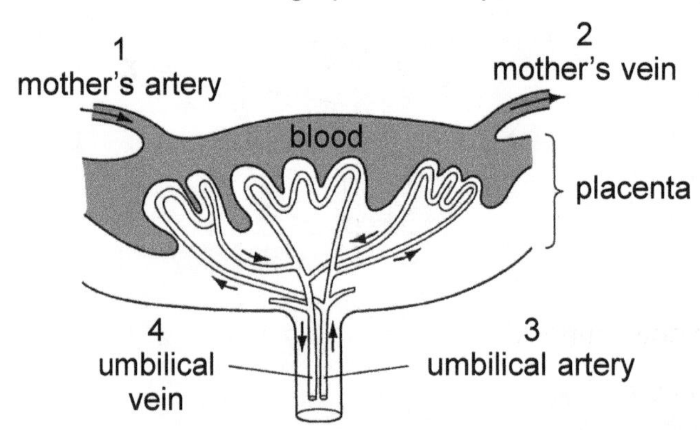

Which parts contain blood with the most oxygen and nutrients?

A 1 and 3
B 1 and 4
C 2 and 3
D 2 and 4

( )
[N10/1/30]

13. Which statement is **not** true of the offspring resulting from asexual reproduction?

A They are produced by self-fertilisation.
B They are produced from a single parent.
C Their cells have the same alleles.
D Their cells have the same number of chromosomes.

( )
[N09/1/27]

14. The following investigation was carried out using flower buds growing on three plants of the same species:

Plant 1 – The anthers were carefully removed and the buds left open to the air.
Plant 2 – The anthers were left untouched and a paper bag was tied tightly around each bud.
Plant 3 – The anthers were carefully removed and a paper bag was tied tightly around each bud.

Although all flowers later opened normally, only those on plant 1 produced seeds. This result shows that in this species

A only cross-pollination can take place.
B only wind-pollination can take place.
C only insect-pollination can take place.
D both self- and cross-pollination can take place.

( )
[N09/1/28]

15. The diagram shows the male reproductive system.

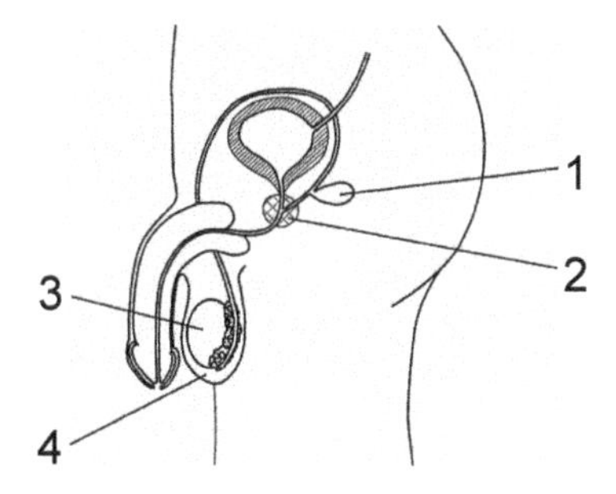

What are the functions of the labelled parts?

	hormone production	seminal fluid production	sperm production
A	1 and 3	3 and 2	3 and 4
B	2 and 3	1 and 2	3 and 4
C	3 only	1 and 2	3 only
D	4 only	2 and 4	3 only

( )
[N09/1/29]

16. Which precautions should be taken to prevent the spread of HIV?

1 avoidance of any direct skin contact with another person
2 medical staff wearing gloves when treating patients
3 not sharing soap used by another person
4 prevent exchange of body fluids being in direct contact
5 treatment of blood products to destroy the virus

A 1, 2 and 3
B 1, 3 and 4
C 2, 3 and 5
D 2, 4 and 5

( )

[N09/1/30]

17. The diagram shows half a flower.

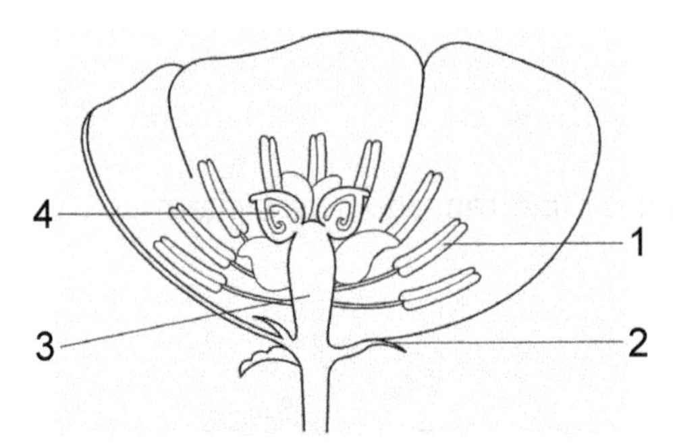

Where are the gametes produced?

A 1 and 3
B 1 and 4
C 2 and 3
D 2 and 4

( )

[N08/1/27]

18. Which features correctly describe a wind-pollinated flower?

	coloured petals	anthers large and hang out of flower	pollen smooth and light	sticky stigma inside flower
A	✓	✓	×	×
B	✓	×	✓	✓
C	×	✓	✓	×
D	×	✓	×	✓

key
✓ = feature present
× = feature absent

( )

[N08/1/28]

19. The diagram shows a developing fetus.

Where does the exchange of oxygen and carbon dioxide between the fetus and its mother take place?

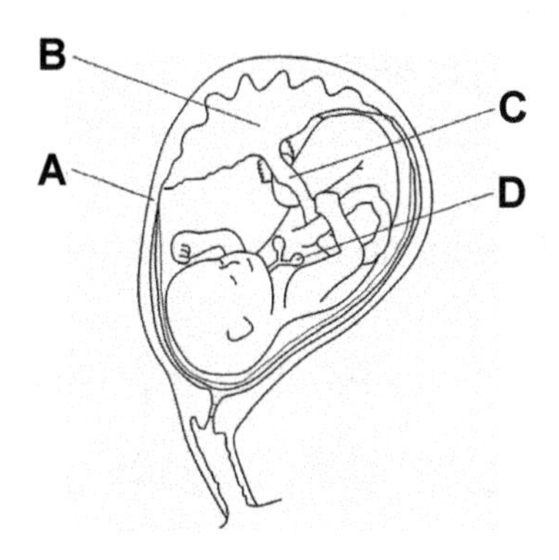

( )

[N08/1/29]

20. The diagram shows a potato plant reproducing asexually by tubers.

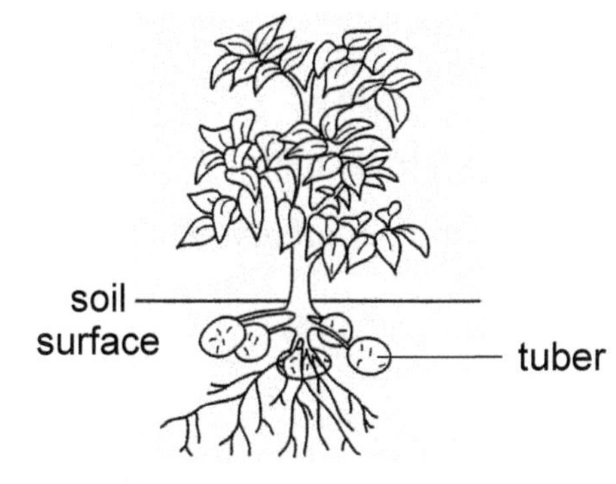

Four observations were made about the potato plant.

1 There is one parent plant.
2 The tubers are attached to the parent.
3 The tubers are genetically identical to the parent.
4 The tubers store food.

Which of these observations describe asexual reproduction?

A 1 and 3
B 1 and 4
C 2 and 3
D 2 and 4

( )

[N07/1/31]

21. Many wind-pollinated flowers have

A feathery stigmas and light pollen.
B feathery stigmas and sticky pollen.
C short stigmas and light pollen.
D short stigmas and sticky pollen. ( )

[N07/1/32]

22. The diagram shows two flowers on plant X and one flower on a different plant Y, of the same species.

Which transfer of pollen will bring about cross pollination?

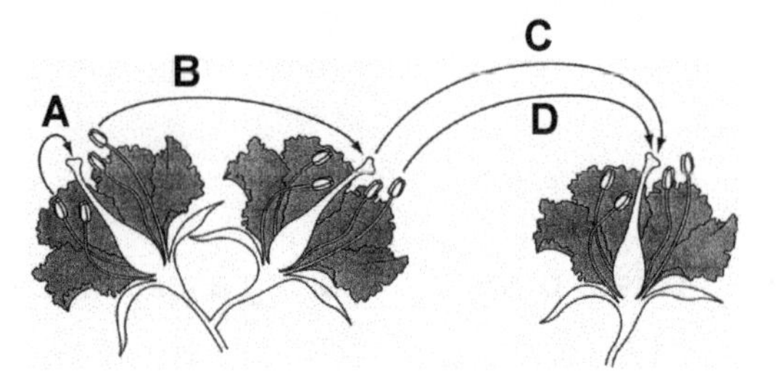

( )

[N07/1/33]

23. The diagram shows part of the female urino-genital system.

Where are sperm deposited during intercourse?

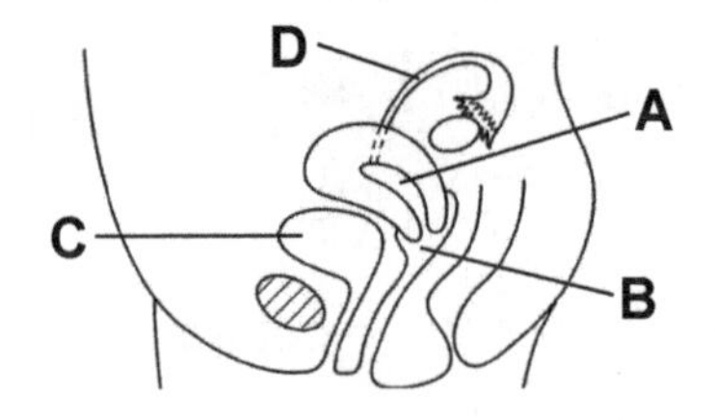

( )

[N07/1/34]

24. What would be the result of cutting the sperm ducts on the right and left sides in a male animal?

A Male sex hormones would no longer circulate in the blood.
B The animal would be unable to pass urine.
C The animal would be unable to develop spermatozoa.
D The animal would become sterile.

( )

[N07/1/35]

25. The diagram shows a developing fetus.

Where does gaseous exchange between mother and fetus occur?

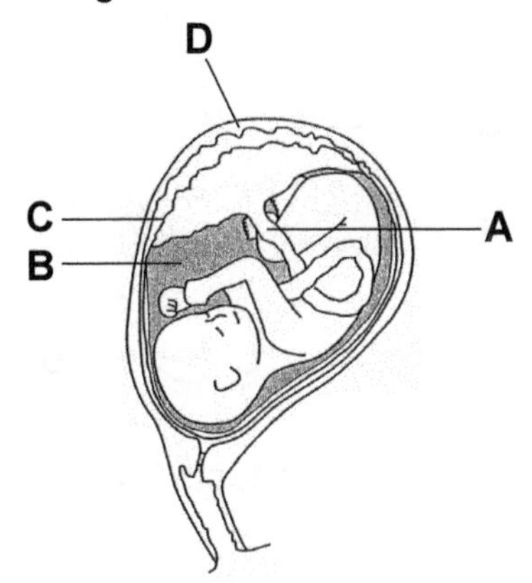

( )

[N06/1/37]

26. The diagram shows the carpel of a flower soon after pollination.

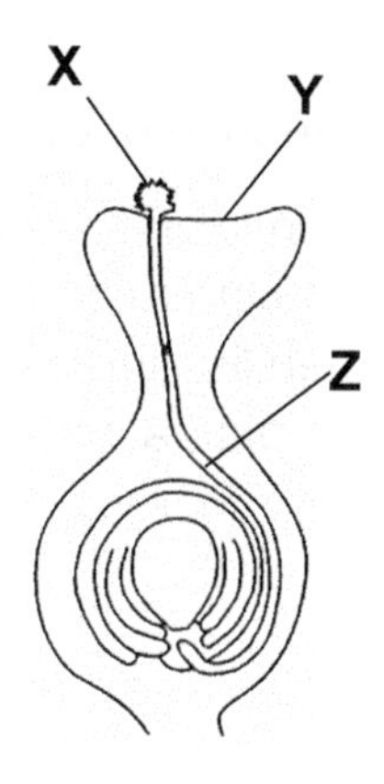

What are the labelled structures?

	X	Y	Z
A	pollen grain	stigma	pollen tube
B	pollen tube	pollen grain	stigma
C	pollen tube	stigma	pollen grain
D	stigma	pollen tube	pollen grain

( )

[N05/1/33]

27. The diagram shows a fetus developing in a uterus.

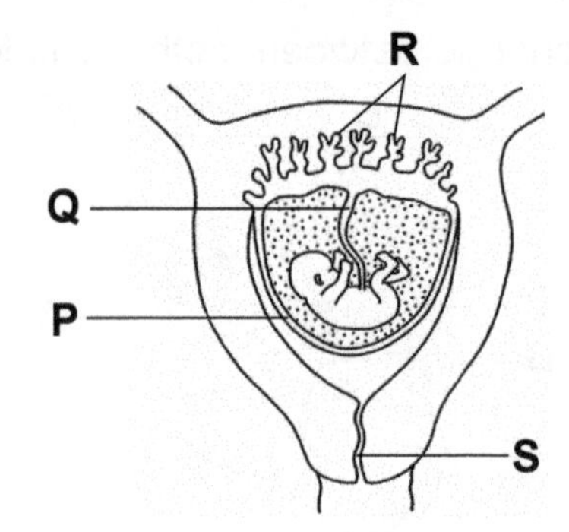

Which structures remove excretory products from the fetus?

A **P** and **Q**
B **Q** and **R**
C **R** and **S**
D **S** and **P**

( )
[N05/1/36]

28. In human reproduction, which sequence of events is correct?

A menstruation → ovulation → fertilisation → implantation
B menstruation → ovulation → implantation → fertilisation
C ovulation → menstruation → fertilisation → implantation
D ovulation → menstruation → implantation → fertilisation

( )
[N05/1/37]

STRUCTURED QUESTIONS . . . .

1. The figure shows a developing fetus in the uterus.

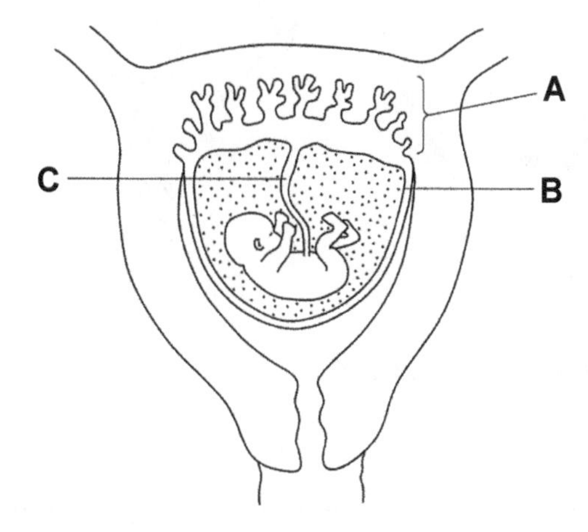

(a) Name the structures labelled **A**, **B** and **C**.

A...

....

B...

....

C...

....

[3]

(b) State **one** function of the amniotic fluid.

....

....[1]

(c) Name **two** substances that pass from the mother to the fetus during pregnancy.

1. ....

....

2. ....

....

[2]

[N11/2/5]

2. The figure shows vertical sections of two flowers A and B.

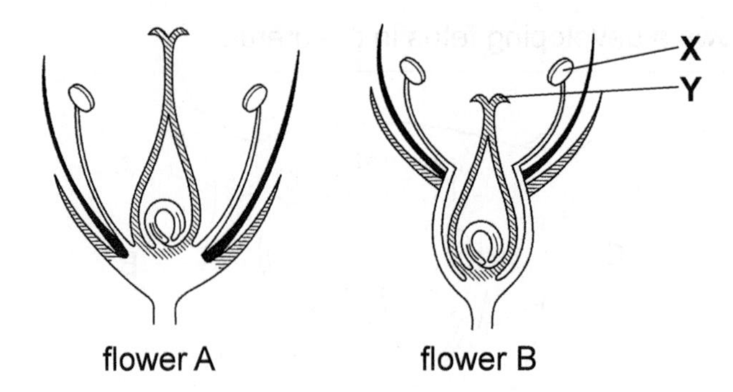

(a) Name the parts labelled **X** and **Y**.

X..

....

Y..

.... [2]

(b) With reference to the figure, state two differences between flower A and flower B.

1. ....

....

2. ....

.... [2]

(c) Define the term *pollination*.

....

....

....[2]

[N11/2/7]

3. The figure shows some of the changes during a menstrual cycle.

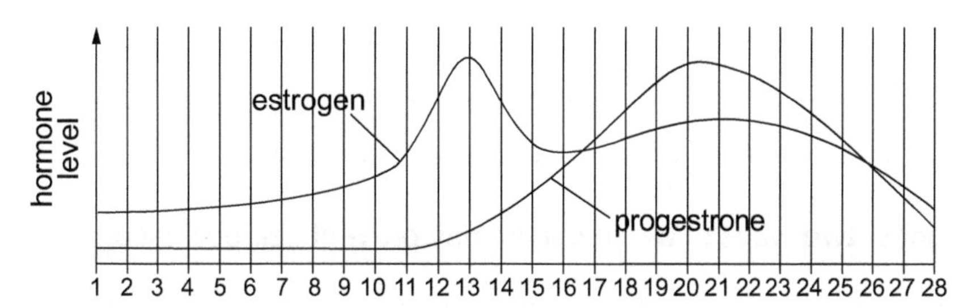

(a) With reference to the figure, describe the changes in the levels of estrogen and progesterone between day 5 and day 14.

estrogen ....

....

....

progesterone ....

....

....[2]

(b) State the effect of the hormone progesterone.

....

....[1]

(c) State what happens to the level of progesterone if fertilisation occurs.

....

....[1]

(d) Outline the early development of a fertilised egg.

....

....

....

....[3]

[N10/2/2]

4. **(a)** Define the term *pollination*.

....

....

....[2]

(b) Suggest **two** ways in which plants may prevent self-pollination.

....

....

....[2]

The figure shows a section of a flower.

(c) Suggest how insect pollination takes place in this flower.

....

....

....[2]

(d) Name the process that follows pollination.

....

....[1]

[N09/2/4]

5. The figure shows a section of a wind-pollinated flower.

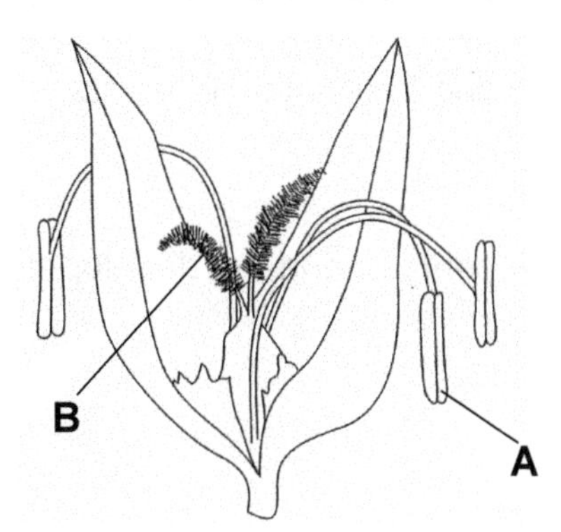

(a) Identify the parts labelled **A** and **B**.

A...

....

B...

....

[2]

(b) State two visible features which indicate that this flower is pollinated by the wind.

1. ....

....

2. ....

....

[2]

(c) Suggest two features of the pollen produced by this flower.

1. ....

....

2. ....

....

[2]

[N08/2/7]

FREE RESPONSE QUESTIONS. . . . .

1. Define the term pollination and fertilisation in plants. [4]

[N06/2/8 Either (a)]

2. Describe the role of the placenta during the development of the fetus. [4]

[N06/2/8 Or (a)]

TOPIC 13
Cell Division

MULTIPLE CHOICE QUESTIONS . . . .

1. The mass of DNA in the nucleus of a diploid body is represented by 2D.

How much DNA will there be in a gamete and in the zygote?

	DNA in gamete	DNA in zygote
A	D	2D
B	2D	4D
C	½D	D
D	2D	2D

( )

[N12/1/31]

2. To which of the processes shown does mitosis contribute?

	genetic variation	increase in cell number	replacement of damaged cells
A	✓	✓	×
B	✓	×	×
C	×	✓	✓
D	×	×	✓

key
✓ = contributes to process
× = does not contribute to process

( )

[N12/1/32]

3. A cell that contains three pairs of homologous chromosomes divides by meiosis.

Which diagram shows the cell in prophase II?

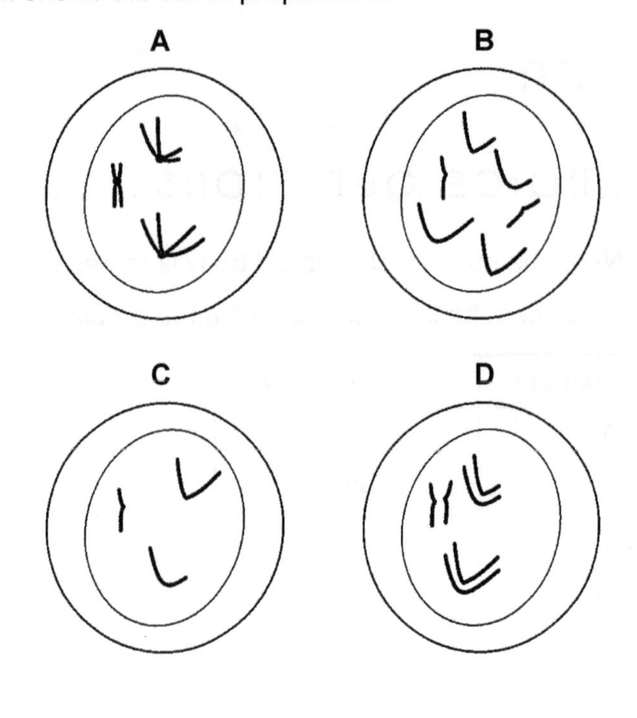

( )

[N12/1/33]

4. Which diagram represents a cell during anaphase of mitosis?

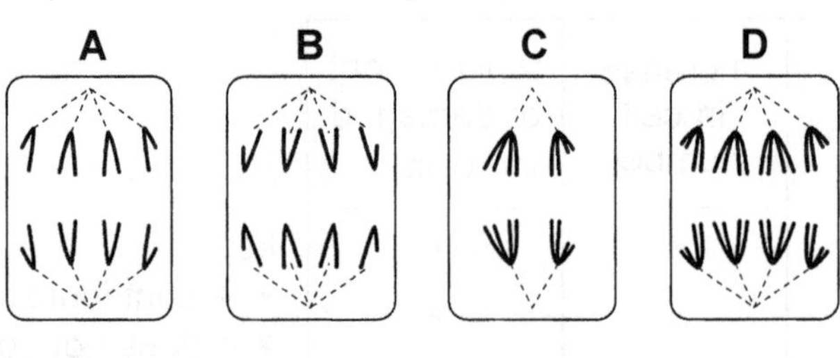

( )

[N11/1/31]

5. The list gives some of the stages involved in gamete and zygote formation.

1 prophase I of meiosis
2 prophase II of meiosis
3 metaphase I of meiosis
4 fertilisation

During which stages do events occur that increase genetic variation in the zygote?

A 1, 2 and 3
B 1, 3 and 4
C 2 and 3 only
D 3 and 4 only

( )

[N11/1/32]
[N08/1/33]

6. What describes homologous chromosomes?

A two chromatids that are joined together to form one chromosome
B two chromatids that have identical alleles
C two chromosomes that have identical alleles
D two chromosomes that form a pair at the start of meiosis ( )

[N11/1/33]

7. Where does genetic variation occur in the production of offspring?

1 prophase 1 meiosis
2 metaphase 1 meiosis
3 prophase 2 meiosis
4 fertilisation

A 1, 2 and 4
B 1, 3 and 4
C 2 and 3 only
D 2 and 4 only ( )

[N10/1/31]

8. Organs may have haploid or diploid cells or both.

Which row is correct?

	organ	diploid cells present	haploid cells present
A	brain	yes	yes
B	liver	no	yes
C	ovary	yes	yes
D	testis	yes	no

( )

[N10/1/32]

9. What produces identical cells as a result of mitosis?

A Half the chromosome number goes into each of the daughter cells.
B The cytoplasm is exactly halved to form the two daughter cells.
C The DNA is exactly copied into the two daughter cells.
D The organelles are equally divided between the two daughter cells. ( )

[N10/1/33]

10. In which structure are the cells dividing by mitosis **only**?

A ovary
B root tip
C stamen
D testis

( )

[N09/1/31]

11. The diagram shows a cell during mitosis.

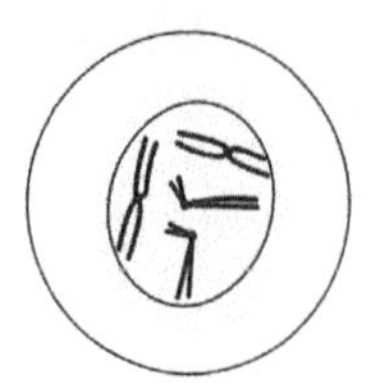

At the end of mitosis what daughter cells will be produced?

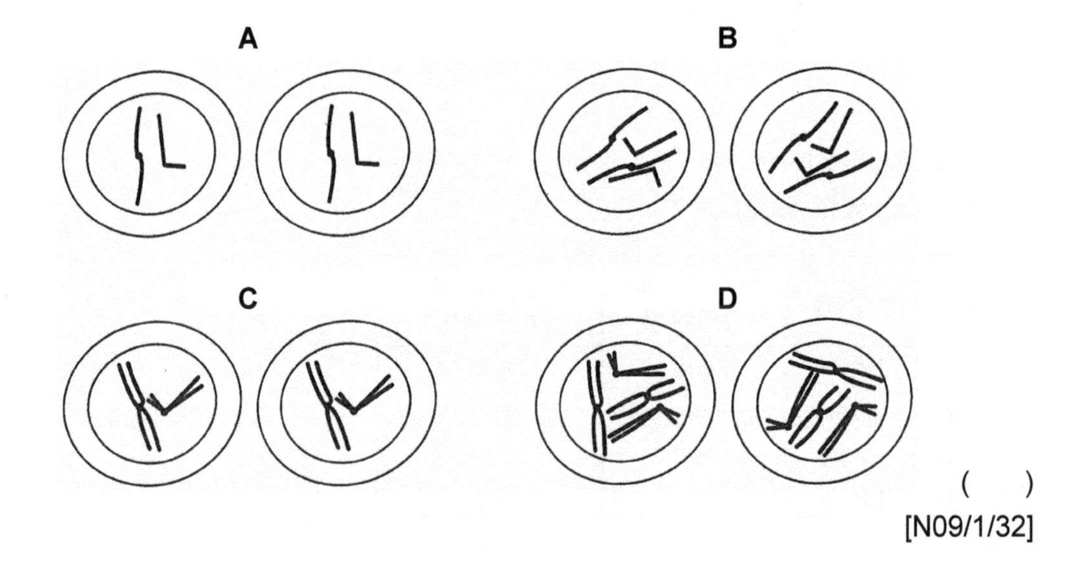

( )

[N09/1/32]

12. The diagram shows the production of a zygote.

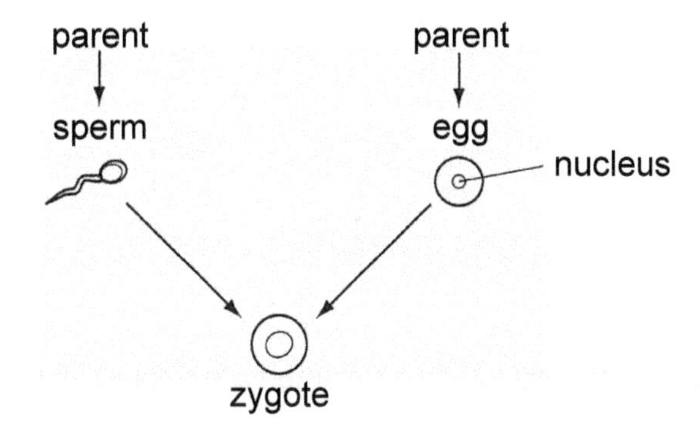

What describes the zygote?

	homologous chromosomes present	produced by meiosis	genetically different from either parent
A	×	✓	✓
B	✓	×	×
C	×	✓	×
D	✓	×	✓

key
✓ = correct
× = incorrect

( )

[N09/1/33]

13. Cell **L** is the product of two cells, **J** and **K**, fusing. Cell **L** undergoes cell division and growth to form an organism. After further development of the organism, cell **V** divides and produces cells **W**, **X**, **Y** and **Z**.

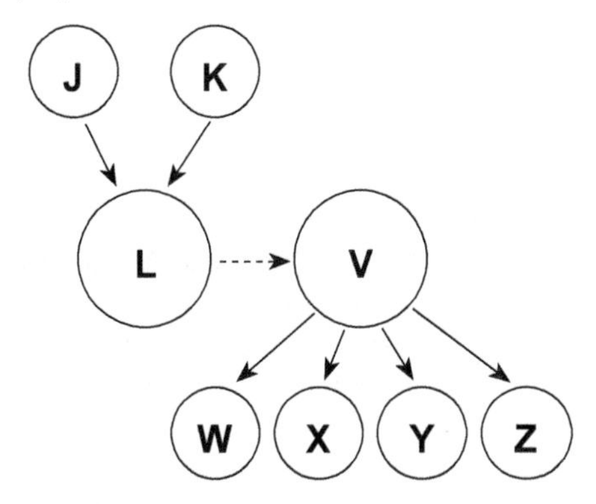

Which two cells are genetically identical and which cell is a zygote?

	genetically identical	zygote
A	**J** and **K**	**L**
B	**J** and **L**	**V**
C	**L** and **V**	**L**
D	**W** and **X**	**V**

( )

[N08/1/30]

14. The diagram shows chromosomes during mitosis.

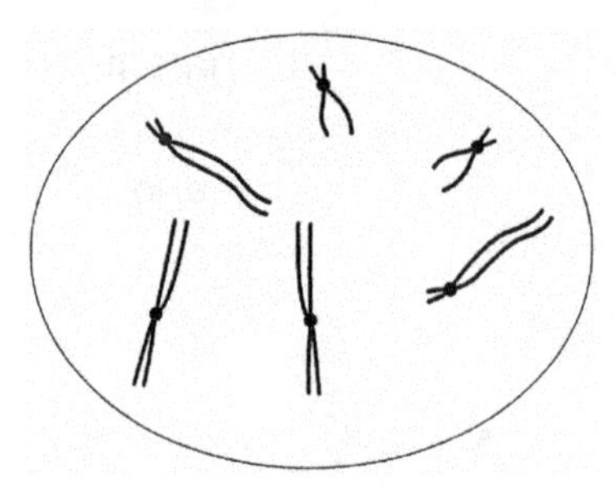

How many pairs of homologous chromosomes are shown and which stage of mitosis is shown?

	pairs of homologous chromosomes	stage of mitosis
A	3	prophase
B	3	telophase
C	6	prophase
D	6	telophase

( )

[N08/1/31]

15. There are 20 chromosomes in each leaf cell of a maize plant.

What is the number of chromosomes in the male nucleus of a pollen grain of maize?

A 5
B 10
C 20
D 40

( )

[N08/1/32]

16. Which line in the table shows the number of sex chromosomes and the other chromosomes in the body cells of a normal adult human female?

	sex chromosomes	other chromosomes
A	X	23
B	XX	22
C	XX	44
D	XX	46

( )

[N07/1/38]

17. A plant has 20 chromosomes in its leaf cells. The plant reproduces both sexually and asexually.

What is the correct number of chromosomes in the gametes and in cells used for asexual reproduction?

	number of chromosomes	
	gametes	cells used for asexual reproduction
A	10	10
B	10	20
C	20	10
D	20	20

( )

[N06/1/34]

18. The skin cells of an animal contain 8 chromosomes.

How many chromosomes will be present in each of the gametes produced by this animal?

A 16
B 8
C 4
D 2

( )

[N05/1/34]

STRUCTURED QUESTIONS . . . .

1. **(a)** The figure shows the chromosomes in an animal cell.

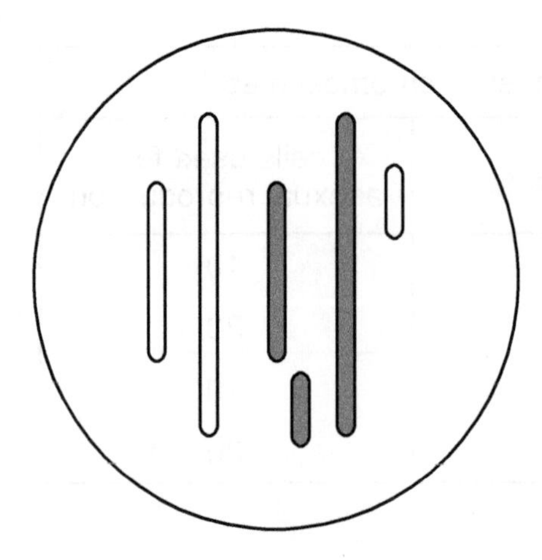

In the spaces below draw diagrams to show the chromosomes of **two** gametes which may be formed when this cell divides by meiosis. [2]

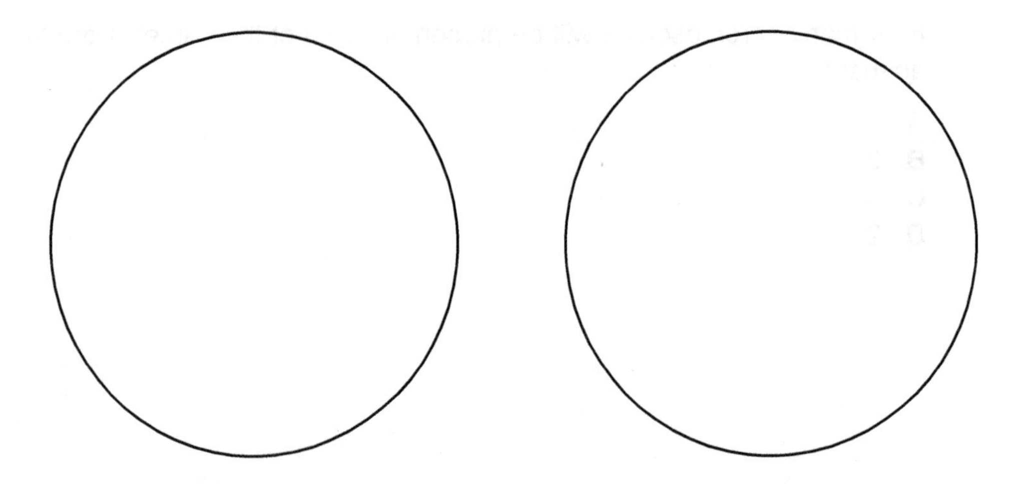

(b) State **one** place in the body where meiosis takes place.

....[1]

[N12/2/7(b)]

2. The figure shows the stages in the human life-cycle.

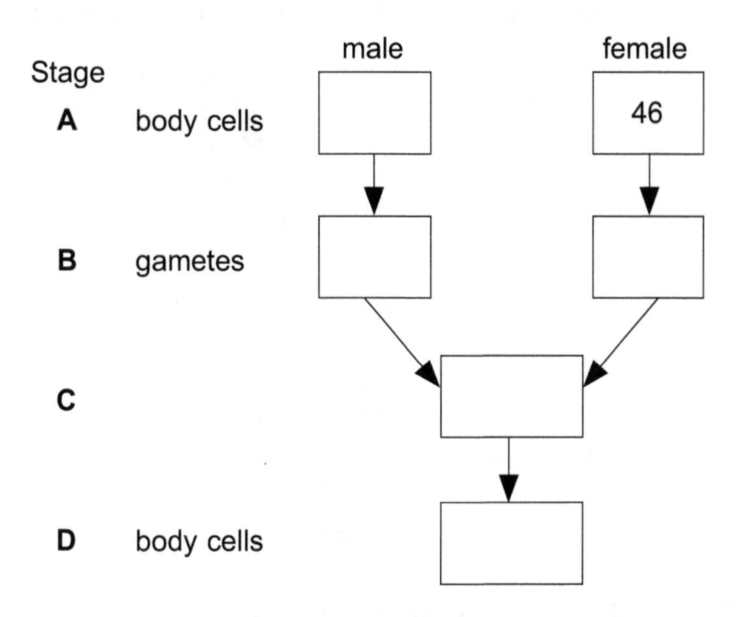

(a) (i) Complete the diagram by writing the number of chromosomes in the boxes representing each stage, **A**, **B**, **C** and **D** of the life-cycle.

The number of chromosomes in the body cells of a female has been completed for you. [4]

(ii) Name the structure at stage **C**.

....

....[1]

(iii) Indicate on the appropriate arrow with the letter **Q** where meiosis occurs in the life-cycle. [1]

(iv) Indicate on the appropriate arrow with the letter **R** where mitosis occurs in the life-cycle. [1]

(b) Define the term *fertilisation*.

....

....

....[2]

[N09/2/1]

3. **Figs. (a)** and **(b)** show the use of some scientific techniques in the process of reproduction. The animal used in this particular procedure was a sheep.

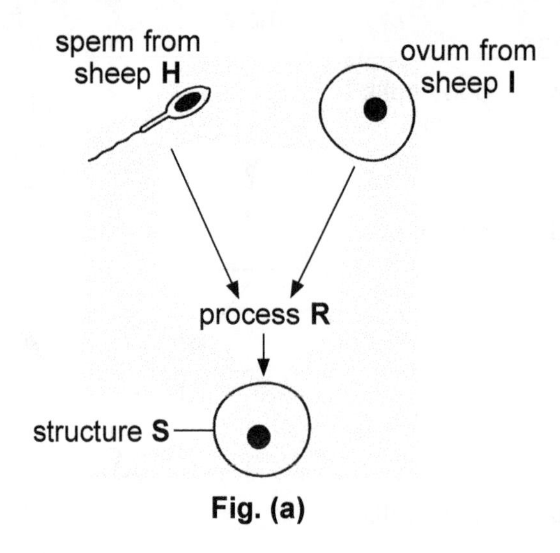

Fig. (a)

(a) Identify process **R** and structure **S** on **Fig. (a)**.

R..

....

S..

....

[2]

(b) The diploid number of chromosomes for a sheep is 54. State the number of chromosomes in

(i) the sperm of sheep **H**, ... [1]

(ii) structure **S**. .... [1]

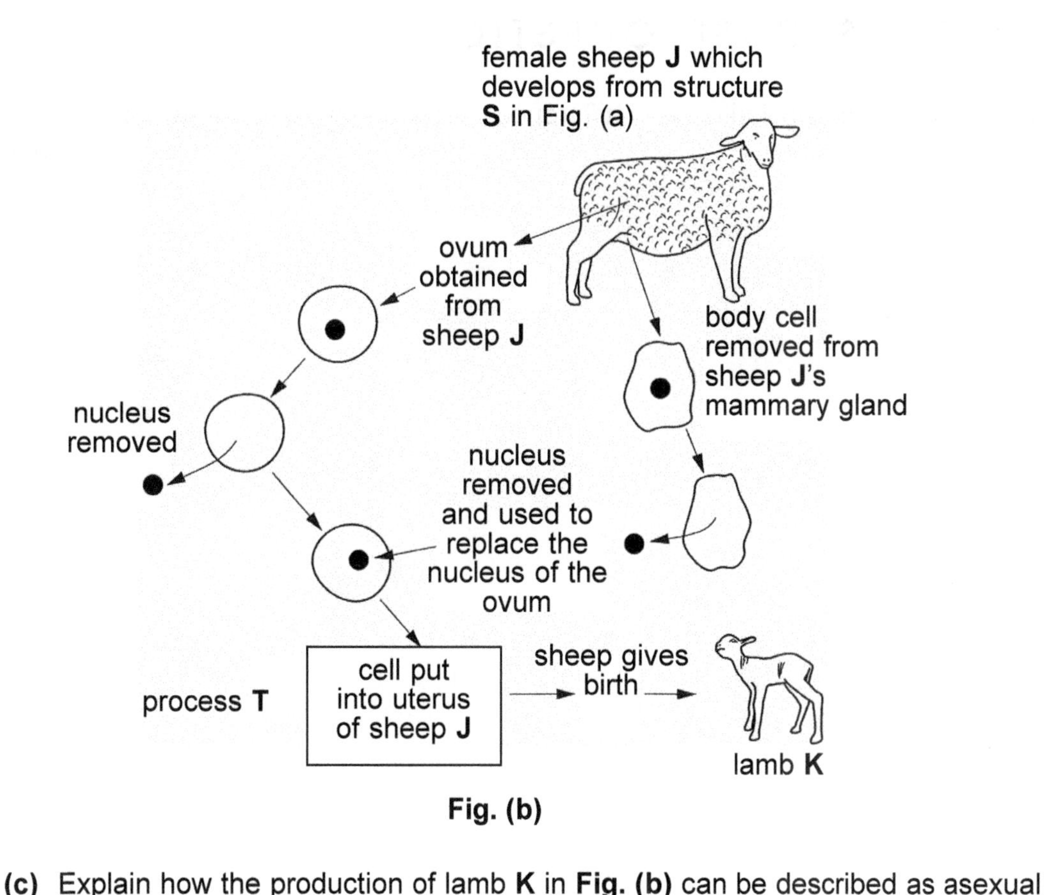

Fig. (b)

(c) Explain how the production of lamb **K** in **Fig. (b)** can be described as asexual reproduction.

....

....[2]

(d) Sex in sheep is inherited in the same way as in humans. With reference to the sex chromosomes, suggest and explain the sex of the lamb **K**.

....

....[2]

(e) It is common, during process **T**, to put the cell into the uterus of a different sheep from the one used to supply the ovum. There are various potential advantages of this procedure. Suggest **one** such advantage.

....

....[1]

[N05/2/4]

FREE RESPONSE QUESTIONS. . . . .

1. Describe the similarities and differences between mitosis and meiosis. [3]

[N10/2/9(a)]

TOPIC 14

Molecular Genetics

MULTIPLE CHOICE QUESTIONS . . . .

1. The table shows the results of mapping 100 nucleotides on a single strand of DNA.

nucleotide	quantity
adenine	22
cytosine	20
guanine	47
thymine	11

How many thymine nucleotides will there be on the strand of DNA that is complementary to this strand?

A 33
B 22
C 20
D 11 ( )

[N12/1/34]

2. Which statement about DNA is correct?

A A molecule of DNA contains many genes.
B A molecule of DNA is larger than a chromosome.
C A molecule of DNA is the same size as a gene.
D Each molecule of DNA is always the same length. ( )

[N12/1/35]

3. Which statement applies **only** to a gene?

A It can be copied during cell division.
B It can control multiple characteristics.
C It codes for a single polypeptide.
D It is composed of four bases. ( )

[N11/1/34]

4. The diagram outlines part of the process to produce recombinant DNA that will synthesise human insulin.
At steps 1, 2 and 3, enzymes have to be used.

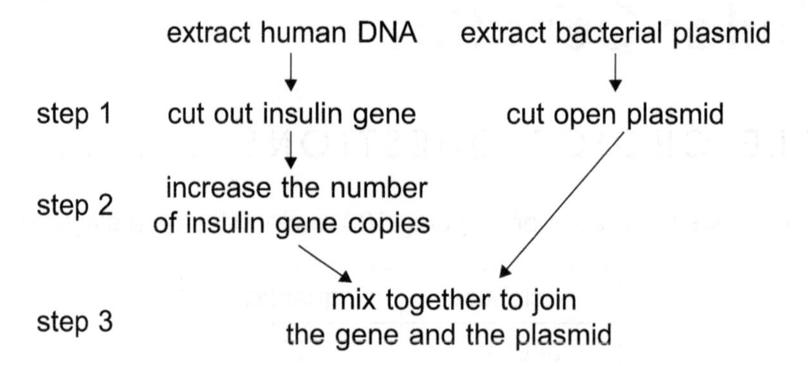

Which row correctly identifies the enzyme in each step?

	step 1	step 2	step 3
A	polymerase	ligase	restriction
B	polymerase	restriction	ligase
C	restriction	ligase	polymerase
D	restriction	polymerase	ligase

( )
[N11/1/35]

5. Which statement describes a gene?

A a base with a sugar and a phosphate group
B a number of DNA molecules
C a sequence of nucleotides
D the chain of alleles on a chromosome

( )
[N10/1/34]

6. Some statements about the production of human insulin from genetically modified bacteria are listed.

1 A bacterial plasmid is cut open.
2 The bacterium is grown in a fermenter.
3 The insulin gene is cut out from human DNA.
4 The insulin gene is inserted into a plasmid.
5 The plasmid is inserted into a bacterium.

What is the correct sequence of these statements?

A $2 \rightarrow 3 \rightarrow 5 \rightarrow 4 \rightarrow 1$
B $2 \rightarrow 5 \rightarrow 3 \rightarrow 1 \rightarrow 4$
C $3 \rightarrow 1 \rightarrow 4 \rightarrow 5 \rightarrow 2$
D $3 \rightarrow 1 \rightarrow 5 \rightarrow 4 \rightarrow 2$

( )
[N10/1/35]

7. A gene contains 900 phosphate groups.

How many bases will it contain?

A 300
B 900
C 1800
D 2700

( )

[N09/1/34]

8. Some stages in the transfer of genes from one species to a second species are listed.

1 Bacteria containing the recombinant DNA replicate.
2 DNA is cut into pieces to isolate the target gene.
3 Organisms from a second species are infected with bacteria.
4 Plasmids from bacteria accept the target gene.

In which order are they performed before the gene is incorporated into the second species?

A $1 \rightarrow 3 \rightarrow 2 \rightarrow 4$
B $2 \rightarrow 4 \rightarrow 3 \rightarrow 1$
C $2 \rightarrow 4 \rightarrow 1 \rightarrow 3$
D $3 \rightarrow 1 \rightarrow 4 \rightarrow 2$

( )

[N09/1/35]

9. What is the correct arrangement for the components in a nucleotide?

A

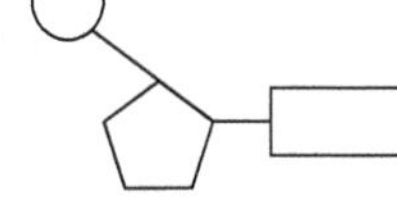

key

sugar

base

B

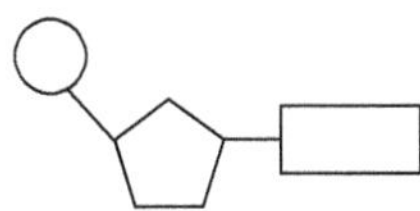

C

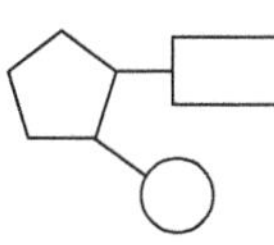

D

( )

[N08/1/34]

10. The diagram shows a process by which a human insulin gene can be inserted into bacterial DNA to produce human insulin.

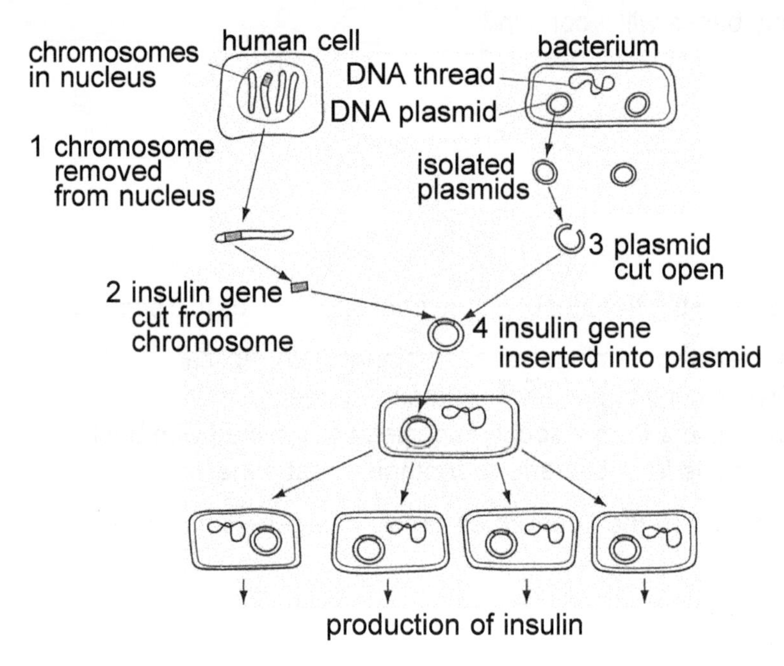

Which stages use a restriction enzyme?

A 1 and 3
B 2 and 3
C 1 and 4
D 2 and 4

( )
[N08/1/35]

FREE RESPONSE QUESTIONS.................

1. Describe the structure and function of DNA. [5]

[N10/2/9(b)]

2. **(a)** Outline the relationship between DNA, genes and chromosomes. [2]

(b) State the significance of the order of bases in DNA molecule. [2]

(c) Describe how bacteria can be used to produce human insulin. [5]

[N09/2/10 Either]

TOPIC 15
Inheritance

MULTIPLE CHOICE QUESTIONS . . . .

1. Which statement about the inheritance of ABO blood groups is correct?

A Allele I^A is dominant to allele I^B.
B Allele I^O is dominant to allele I^A.
C Allele I^O is recessive to allele I^B.
D Alleles I^B and I^O are co-dominant. ( )

[N12/1/36]

2. Which is correct for the number of chromosomes and type of sex chromosomes in a body cell of a human male?

	total number of chromosomes present	type of sex chromosomes present
A	44	XY
B	44	Y
C	46	XY
D	46	Y

( )

[N12/1/37]

3. In terms of natural selection, what is the significance of the survival of the fittest in a tiger population?

A They achieve the best physical condition.
B They contribute the most offspring to the next generation.
C They have the most effective camouflage.
D They are the most sexually active. ( )

[N11/1/36]

4. The inheritance pattern of an abnormal condition in four families is shown.
Which family proves that the condition must be caused by a dominant allele?

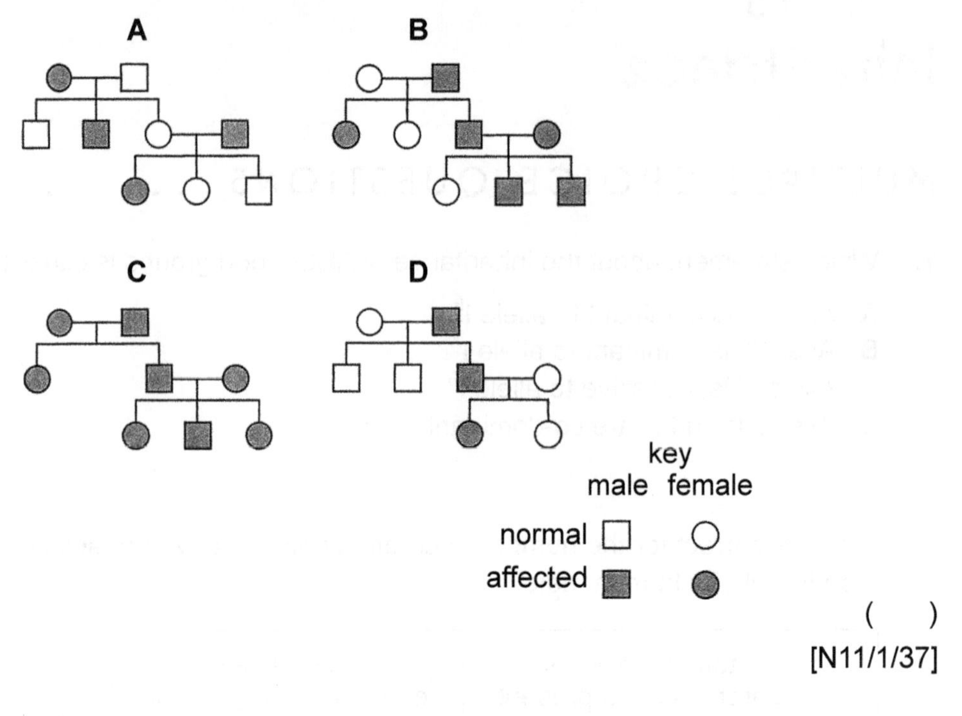

( )

[N11/1/37]

5. Which two statements about continuous variation are correct?

1 The heights of adult humans will partly depend on the quality of their diets when young.
2 The faster period of growth in humans is in the embryo.
3 A group of adult males had heights ranging from 155 cm to 220 cm.
4 Humans have stopped growing by the time they are 22 years old.
5 Humans grow taller during babyhood and childhood.

A 1 and 2
B 1 and 3
C 2 and 4
D 3 and 5

( )

[N11/1/38]

6. Which one of these structures is part of the other three?

A chromosome
B gamete
C gene
D nucleus

( )

[N10/1/36]

7. In rabbits, brown fur is dominant to white fur.

Two rabbits, both heterozygous for fur colour, are mated and a litter is produced. The first three rabbits to be born have brown fur.

What is the chance of the fourth one having white fur?

A 25%
B 50%
C 75%
D 100%

( )

[N10/1/37]

8. Which statement about the evolution of life on Earth is correct?

A Artificial and natural selection have both contributed towards evolution.
B Artificial selection has removed harmful mutations during evolution.
C Evolution can only be explained by natural selection.
D Mutations have increased variation on which natural selection has worked.

( )

[N10/1/38]

9. In mice, the normal brown fur colour is dominant to white fur.

A girl mated two mice that she knew were both heterozygous for fur colour. When the litter of mice were being born, the first three mice all had brown fur. The girl then said that the next mouse to be born would have white fur.

What are the chances that she would be correct?

A 25%
B 50%
C 75%
D 100%

( )

[N09/1/36]

10. Which fertilisation would result in a child with Down's Syndrome?

	chromosomes in ovum	chromosomes in sperm
A	23	23
B	24	24
C	24	23
D	47	46

( )

[N09/1/37]

11. Which is a correct example of the process named?

	process	example
A	artificial selection	tiger population reduced in a game reserve to protect herbivores
B	conservation	pesticides used to kill weeds in a field of crops to increase food production
C	evolution	mountain tigers develop a thicker layer of fur in response to falling temperatures
D	natural selection	weaker young elephants die due to lack of food, leaving a stronger population

( )

[N09/1/38]

12. Which statement uses genetic terminology and information correctly?

A Recessive alleles are usually harmful and dominant alleles are beneficial.

B When two animals both heterozygous for fur colour, are crossed, their offspring will have the same phenotype.

C When two people, both homozygous for blood group A, have children, both parents and offspring will have the same genotype.

D Under the ABO blood grouping system, alleles I^A, I^B and I^O are co-dominant.

( )

[N08/1/36]

13. Which two characteristics both show discontinuous variation?

A blood group, gender

B eye colour, height

C gender, weight

D weight, eye colour

( )

[N08/1/37]

14. The statements refer to natural selection.

1 Competition between organisms alters their genes.
2 More organisms are produced than reach maturity.
3 Organisms inherit characteristics from their parents.
4 Organisms vary in their adaptations.
5 Only one species can occupy an ecological niche.
6 Well-adapted organisms survive and reproduce.

Which four statements summarise the theory of evolution by natural selection?

A 1, 2, 3 and 5
B 1, 2, 4 and 5
C 2, 3, 4 and 6
D 2, 3, 5 and 6 ( )

[N08/1/38]

15. Which two statements about continuous variation are correct?

1 The heights of adult humans will partly depend on the quality of their diets when young.
2 During puberty there is a dramatic growth spurt.
3 A group of adult males had heights ranging from 155 cm to 220 cm.
4 During old age, people tend to shrink in height.
5 Humans grow taller during babyhood and childhood.

A 1 and 2
B 1 and 3
C 2 and 4
D 3 and 5 ( )

[N07/1/36]

16. Humans can inherit either brown or blue eyes. Two brown-eyed parents have three blue-eyed children.

Which two statements are correct for this situation?

1 Each parent has an allele for brown eyes and an allele for blue eyes.
2 The allele for blue eyes is recessive.
3 The probability that their next child will have blue eyes is 0.75.
4 The probability that their next child will have brown eyes is 0.5.

A 1 and 2
B 1 and 3
C 2 and 4
D 3 and 4 ( )

[N07/1/37]

17. In mice, the allele for black fur colour is dominant to the allele for white fur colour. What does this mean in a mouse population?

A Mice with black fur are more successful breeders.
B Most mice have black fur.
C When a black-furred mouse breeds with a white-furred one, the offspring will have grey fur.
D White-furred mice are only born to two white-furred parents. ( )

[N07/1/39]

18. Albinism is an inherited condition caused by a recessive allele a. A is the dominant allele for the normal condition.

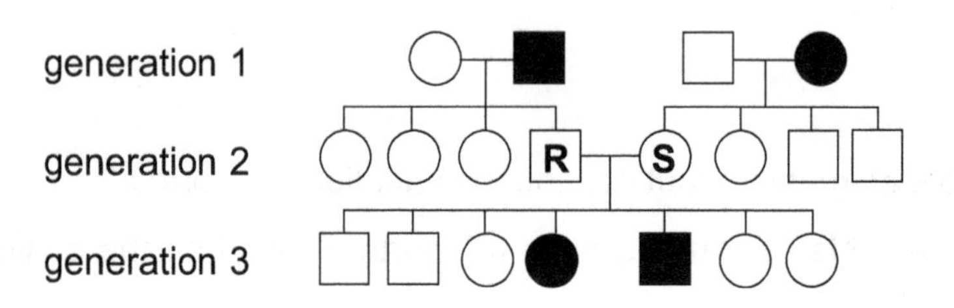

key
normal female
albino female
normal male
albino male

What are the genotypes of individuals **R** and **S**?

	R	S
A	AA	AA
B	AA	Aa
C	Aa	Aa
D	aa	aa

( )

[N07/1/40]

19. Dillip and Shabnam made four statements about themselves.

	Dillip	Shabnam
1	I am a boy.	I am a girl.
2	I am 150 cm tall.	I am 153 cm tall.
3	I am not very good at games.	I am good at games.
4	My blood group is A.	My blood group is AB.

Which statements describe characteristics that show discontinuous variation?

A 1 and 2
B 1 and 4
C 2 and 3
D 3 and 4

( )

[N06/1/38]

20. A recessive homozygous plant is crossed with a dominant homozygous plant and the resulting F_1 plants are self-pollinated.

What will be the phenotypes of the F_2 generation?

A all dominant
B 0.75 dominant 0.25 recessive
C 0.5 dominant 0.5 recessive
D 0.25 dominant 0.5 heterozygous 0.25 recessive

( )

[N06/1/39]

21. A man with blood group I^BI^O and a woman with blood group I^AI^O produce five children.

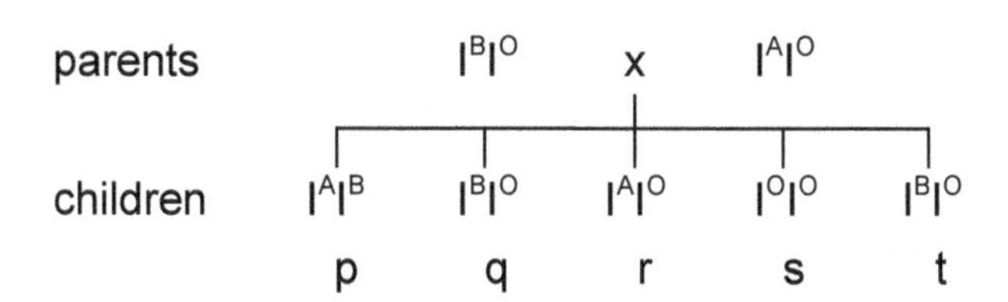

Which child has a blood group that is heterozygous and which child has a blood group that is co-dominant?

	co-dominant	heterozygous
A	p	q
B	p	s
C	t	r
D	s	p

( )

[N06/1/40]

22. In goats, the allele for black hair is dominant to the allele for red hair.

Two black-haired goats mated and produced twelve offspring. Of the first eleven, eight had black hair and three had red hair.

What is the probability of the twelfth offspring having red hair?

A 0.75
B 0.50
C 0.33
D 0.25

( )

[N05/1/35]

23. A woman has blood group O. Her child also has blood group O.

Which blood group can her husband **not** have?

A A
B B
C AB
D O

( )

[N05/1/38]

24. Sex in humans is determined by **X** and **Y** chromosomes inherited from parents.

Which shows the chromosome inherited from the father and from the mother?

	sex of child	chromosome from father	chromosome from mother
A	male	**X**	**Y**
B	male	**Y**	**X**
C	female	**X**	**Y**
D	female	**Y**	**X**

( )

[N05/1/39]

25. Which two characteristics both show continuous variation?

A eye colour and height
B gender and eye colour
C height and weight
D weight and blood group

( )

[N05/1/40]

STRUCTURED QUESTIONS . . . .

1. **(a)** Define the term *gene*.

....

....

.... [2]

(b) Albinism is a condition in which individuals lack the pigment melanin in the skin and iris.

It is caused by a recessive gene, **a**.

The gene for pigmented skin can be represented by **A**.

A family consisted of two parents with pigmented skin, two children with pigmented skin and five albino children.

(i) State the genotype of the parents with pigmented skin.

.... and .... [1]

(ii) State the expected ratio of children with pigmented skin to albino children.

....[1]

(iii) State the possible genotypes of the children with pigmented skin.

....[2]

[N12/2/1]

2. **(a)** **(i)** Define the term *mutation*.

....

....[1]

(ii) State two causes of mutation.

1. ....

....

2. ....

....

[2]

(b) The figure shows the chromosomes of two people, person **A** and person **B**.

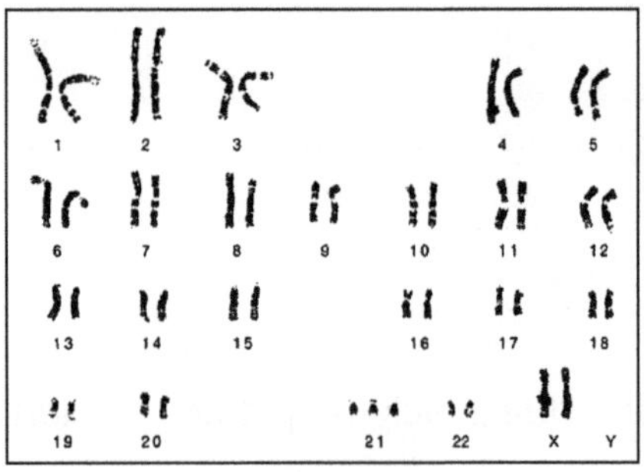

person **A** 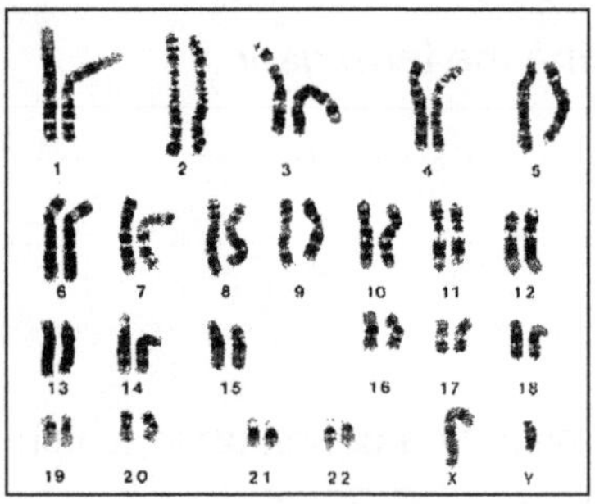 person **B**

(i) State two differences, in terms of numbers and types, between the chromosomes of person **A** and the chromosomes of person **B**.

1. ....

....

2. ....

....

[2]

(ii) Person **A** has inherited a genetic condition.
Name the condition.

....

....[1]

(iii) State the gender of person **B**.
Give a reason for your answer.

....

....

....[2]

[N11/2/2]

3. **(a)** State the type of variation shown by human ABO blood groups.

....

....[1]

(b) Complete the diagram to show the possible blood groups of the children of a father who is heterozygous for blood group A and a mother with blood group AB.

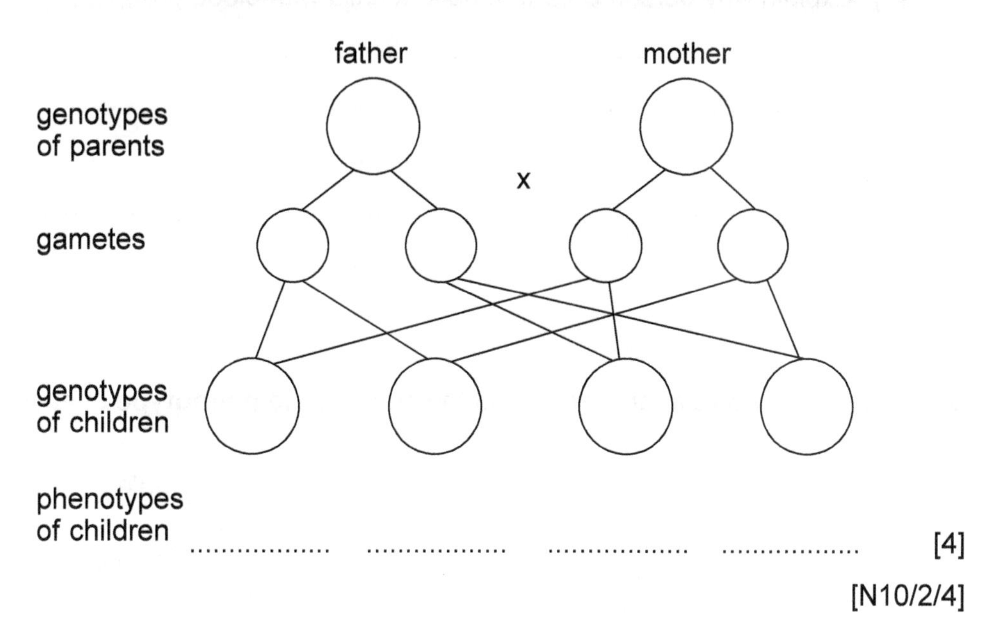

.... [4]

[N10/2/4]

4. The figure shows the blood groups in a family.

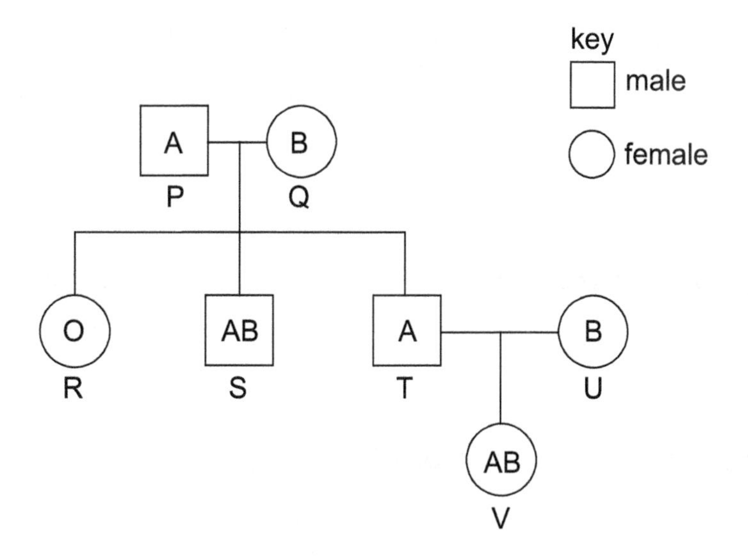

(a) State the genotype of person **P**.

....

....[1]

(b) State the possible genotypes of person **U**.

....

....

....[2]

(c) Explain why person **S** cannot have a child with blood group O.

....

....

....

....

....[2]

[N08/2/4]

5. The figure shows a family tree and the blood group **phenotype** of each individual.

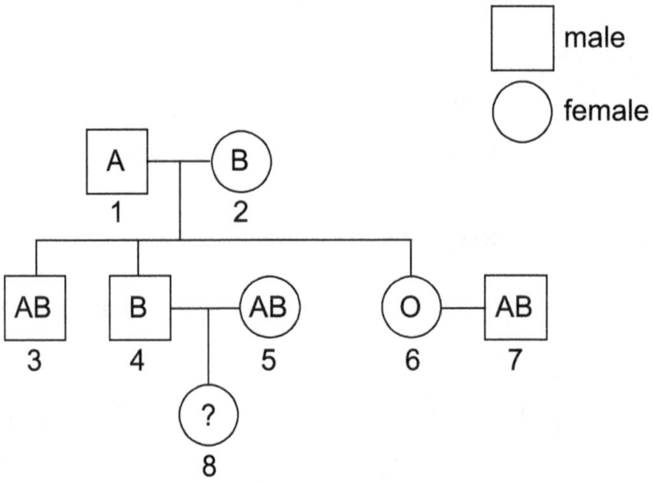

male

female

A 1 — B 2

AB 3, B 4 — AB 5, O 6 — AB 7

? 8

(a) List the alleles involved in the inheritance of ABO blood groups.

....

....[1]

(b) State and explain the **genotype** of individual **1**.

....

....

....

....[3]

(c) Explain why individual **8** cannot have group O blood.

....

....

....

....[2]

(d) State and explain the chances of a child born to individuals **6** and **7** being a boy with group A blood.

....

....

....

....

....[4]

[N07/2/5]

6. Canavan disease is an inherited disease which affects the nervous system.

It is caused by recessive alleles of a gene.

(a) Draw a full genetic diagram to show how a mother and father, who do not have the disease, can have a child with the disease.

Use these symbols:

A for the allele for not suffering from the disease

a for the allele for Canavan disease

[5]

(b) Recessive alleles of genes are formed by mutations. Explain what is meant by:

(i) *a gene* ....

....

....

....[2]

(ii) *a mutation* ....

....

....

....[1]

(c) The table shows the percentage frequency of blood groups in two populations.

	frequency as a percentage of the population			
population	group O	group A	group B	group AB
1	28	27	32	13
2	61	39	0	0

Using only the information in the table, suggest an explanation for the lack of blood group AB in population 2.

....

....

....[2]

[N06/2/2]

TOPIC 16

Organisms and their Environment

MULTIPLE CHOICE QUESTIONS . . . .

1. Changes in the climate may lead to the melting of sea ice and the thawing of ice on the land in the Antarctic that has been frozen for a very long time.

What could lead to evolution in this situation?

A Animals adapt their features to suit the new environment.
B Previously advantageous features may become disadvantageous.
C Seeds dormant for thousands of years could germinate.
D There would be less competition for space to live. ( )

[N12/1/38]

2. The diagram shows how energy from food is used by an animal.

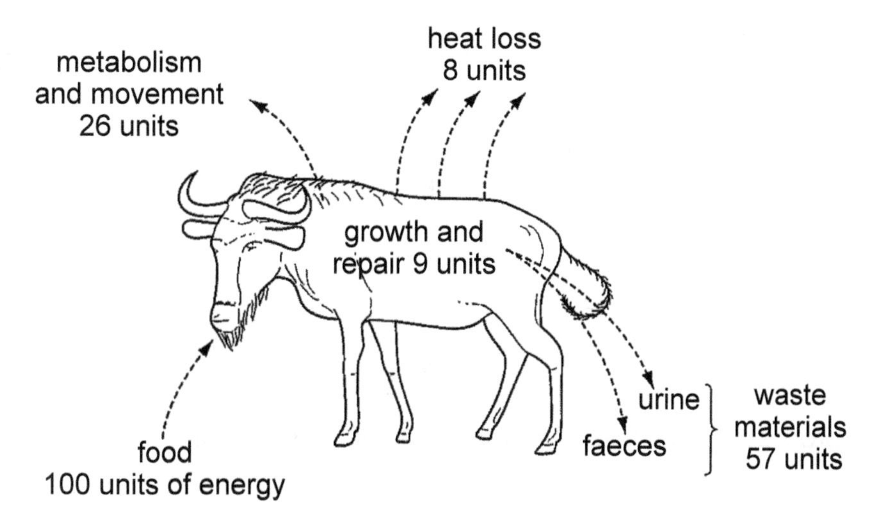

What percentage of this energy is available to consumers and decomposers?

A 91
B 66
C 34
D 9 ( )

[N12/1/39]

3. The diagram represents the carbon cycle. The arrows represent different processes that occur in the carbon cycle.

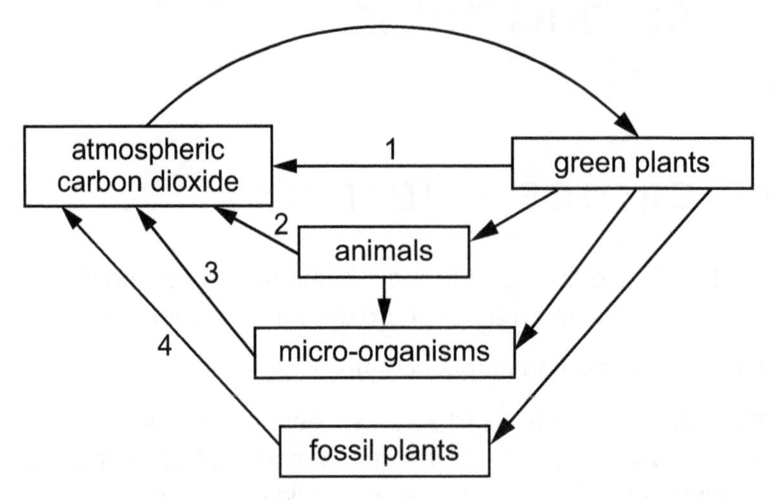

Which of the numbered arrows show where respiration takes place?

A 1, 2, 3 and 4
B 1, 2 and 3 only
C 1 and 4 only
D 2 and 3 only

( )
[N12/1/40]

4. The diagram shows a pyramid of biomass for the food chain:

grass → zebra → lion.

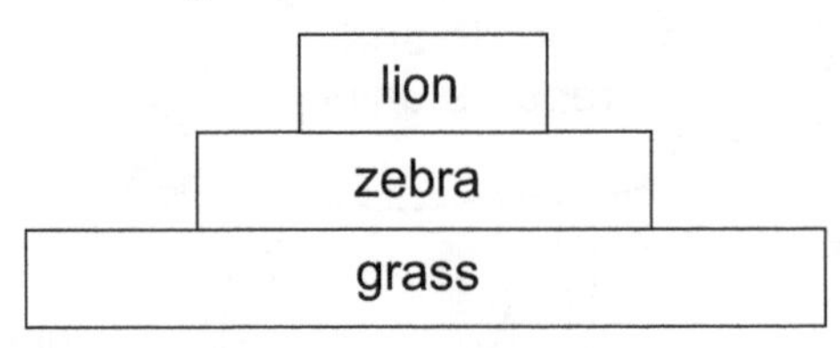

Which process causes loss of biomass from this food chain?

A digestion
B growth
C photosynthesis
D respiration

( )
[N11/1/39]

5. Which group(s) of organisms are **not** necessary for the carbon cycle to continue?

A bacteria and fungi
B green plants and fungi
C green plants only
D herbivores only

( )
[N11/1/40]

6. Which resources are constantly recycled to maintain life?

	carbon	energy	oxygen
A	✓	✓	✓
B	✓	✗	✓
C	✓	✗	✗
D	✗	✓	✗

key
✓ = recycled
✗ = not recycled

( )
[N10/1/39]

7. The diagram shows part of a food web.

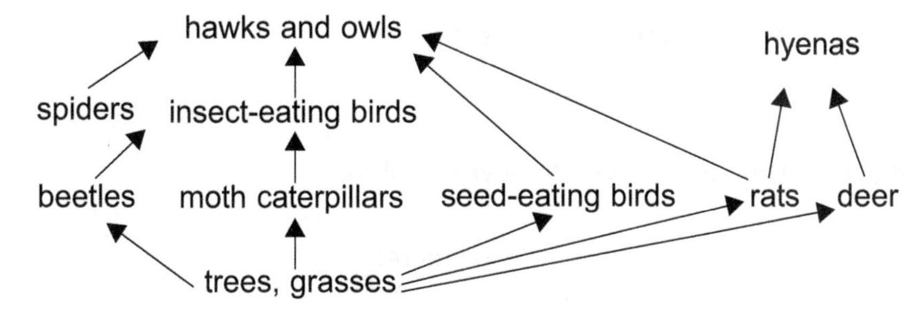

Which organisms are producers, primary consumers and secondary consumers?

	producers	primary consumers	secondary consumers
A	moth caterpillars	insect-eating birds	hawks
B	hawks	seed-eating birds	grasses
C	grasses	spiders	beetles
D	trees	beetles	spiders

( )
[N10/1/40]

8. A crop plant is modified to be resistant to a herbicide.

What is an economic advantage of this modification?

A The crop plant will be resistant to insect pests.
B The crop plant will grow faster and produce a higher yield.
C The herbicide is no longer needed for the production of the crop.
D The herbicide may be used on the crop to kill unwanted weeds. ( )

[N09/1/39]

9. Which microbes are used in the treatment of sewage?

A aerobic bacteria and anaerobic bacteria only
B aerobic bacteria, anaerobic bacteria and fungi
C anaerobic bacteria and viruses
D anaerobic bacteria, fungi and viruses ( )

[N09/1/40]

10. The diagram shows a food web on a wild fruit tree.

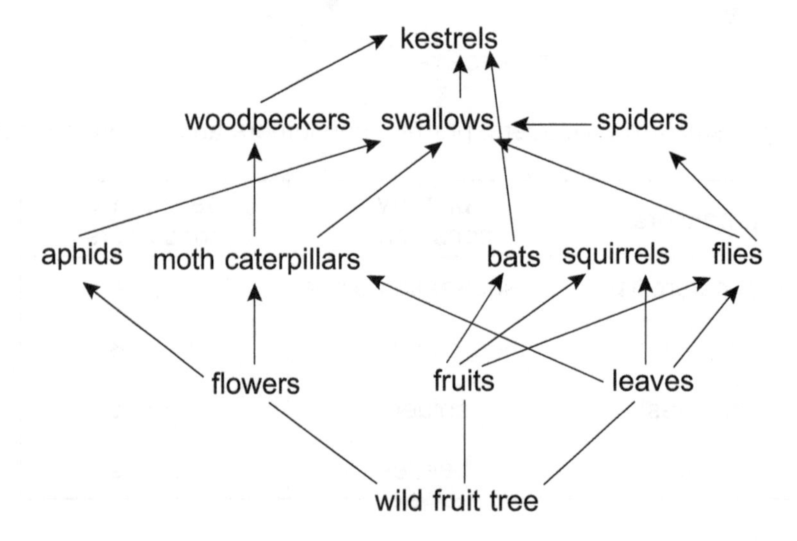

Which animals would be most affected, if the flowers of the tree were **not** pollinated?

A aphids
B bats
C kestrels
D squirrels ( )

[N08/1/39]

11. A farmer sprays insecticide on his crops for a year. The insecticide washes off into a lake where it is absorbed by the producer to enter the food chain.

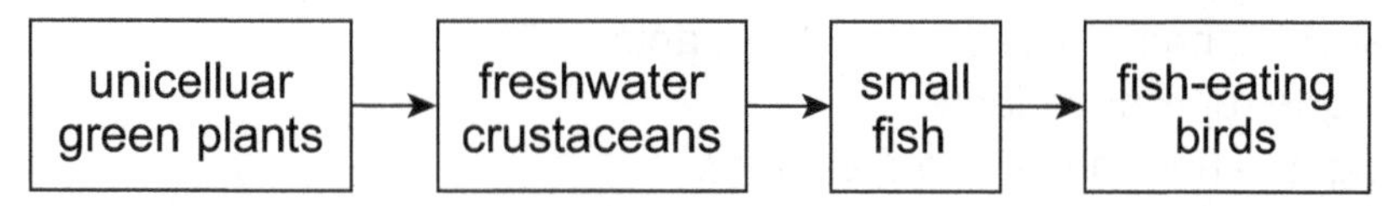

Which gives the levels of insecticide in these organisms at the end of the year? (ppm = parts per million)

	unicellular green plants / ppm	freshwater crustaceans / ppm	small fish / ppm	fish-eating birds / ppm
A	0.05	0.5	0.05	0.05
B	0.05	0.05	0.05	0.05
C	0.05	0.5	5.0	25.0
D	25.0	5.0	0.5	0.05

( )

[N08/1/40]

12. The diagram shows a pyramid of numbers in an ecosystem on land.

Which organisms are smallest in size?

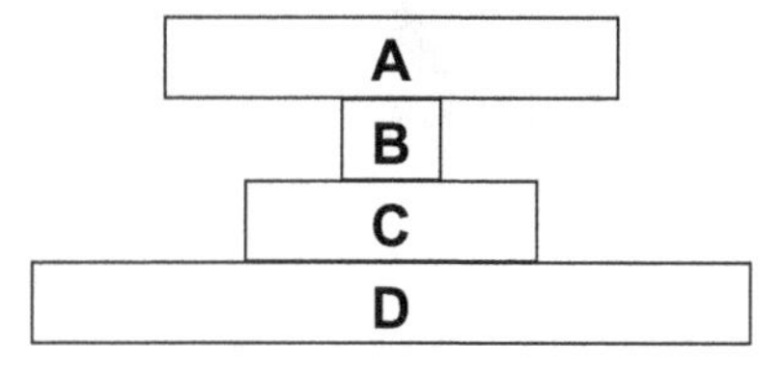

( )

[N07/1/27]

13. The diagram shows four food chains leading to humans.

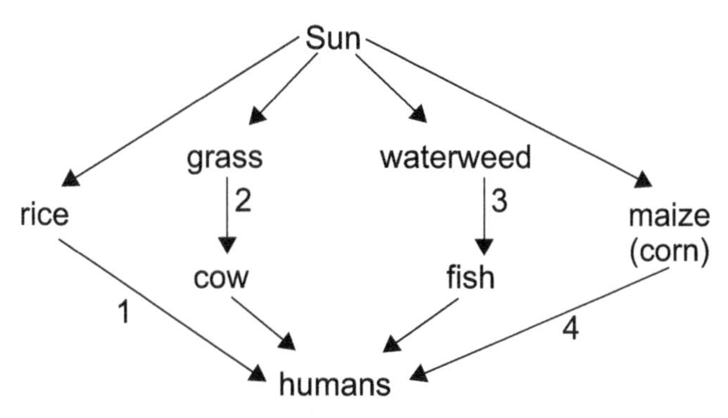

Which chains make the most efficient use of solar energy?

A 1 and 2
B 1 and 4
C 2 and 3
D 2 and 4

( )

[N07/1/28]

14. Inorganic fertiliser is applied each year to fields bordering a lake. The fertiliser runs off into the lake and causes six changes which together make the fish die.

1 Aerobic bacteria feed on dead plants.
2 Algae reproduce faster.
3 Light cannot penetrate the water.
4 Oxygen levels fall.
5 Water becomes green.
6 Underwater plants die.

In which order do the changes take place?

A 2 → 5 → 1 → 6 → 4 → 3
B 2 → 5 → 3 → 6 → 1 → 4
C 3 → 6 → 1 → 4 → 2 → 5
D 4 → 6 → 1 → 2 → 5 → 3

( )
[N07/1/29]

15. The diagram shows part of the carbon cycle.

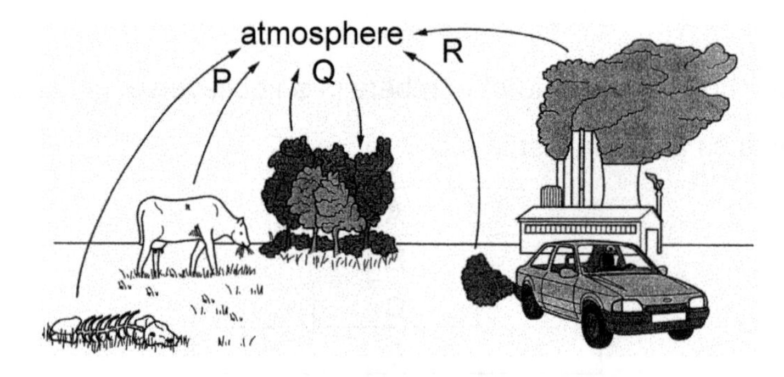

If the use of fossil fuels is reduced, how will this change the levels of carbon dioxide at points P, Q and R?

	P	Q	R
A	decrease	no change	increase
B	increase	decrease	increase
C	no change	decrease	decrease
D	no change	increase	no change

( )
[N07/1/30]

16. Which is a possible sequence for energy flowing through a food web?

	lost as heat	present in glucose	present in protein	recycled for photosynthesis
A	–	2	1	3
B	1	–	3	2
C	2	3	–	1
D	3	1	2	–

( )

[N06/1/29]

17. The diagram shows part of a food web.

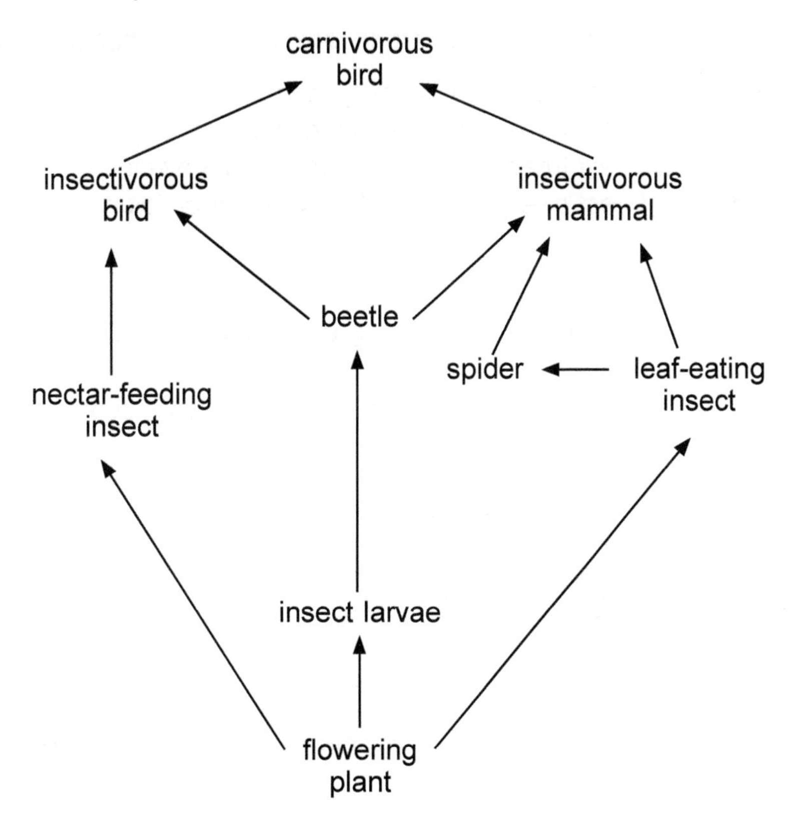

Which is a primary consumer?

A beetle
B carnivorous bird
C insectivorous mammal
D nectar-feeding insect

( )

[N06/1/30]

18. The diagram shows how **carbon** circulates in nature.

Through which stage does most **energy** flow?

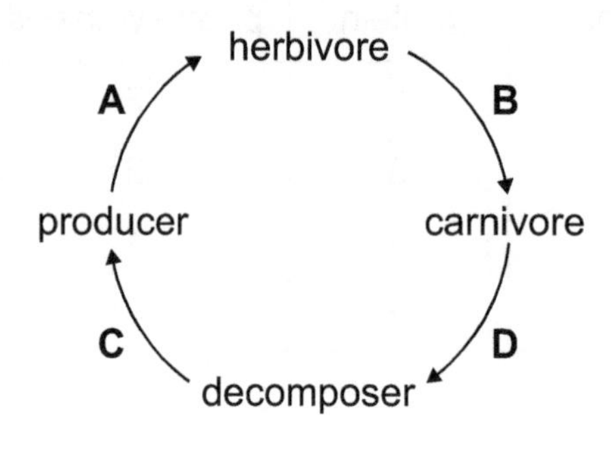

( )

[N06/1/31]

19. What will increase the amount of carbon dioxide in the atmosphere most?

A cutting down forest trees and allowing them to decompose
B cutting down forest trees and leaving the soil bare
C cutting down forest trees and planting crops on the soil
D cutting down forest trees and using the wood for building

( )

[N06/1/33]

20. The diagram shows some animals and green plants sealed in an aquarium.

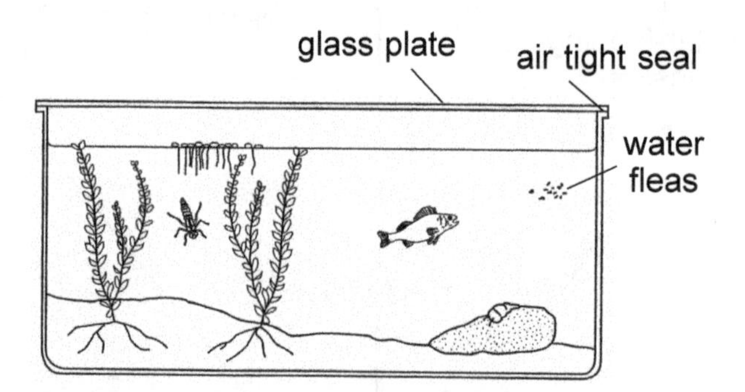

What must be supplied to keep the organisms alive for the longest possible time?

A carbon dioxide
B light energy
C nitrates
D oxygen

( )

[N05/1/27]

21. Some types of bacteria make carbohydrates from carbon dioxide and water using the energy from sunlight.

What are these bacteria?

A anaerobic
B consumers
C decomposers
D producers ( )

[N05/1/30]

22. The diagram shows a forest and farmland on either side of a river.

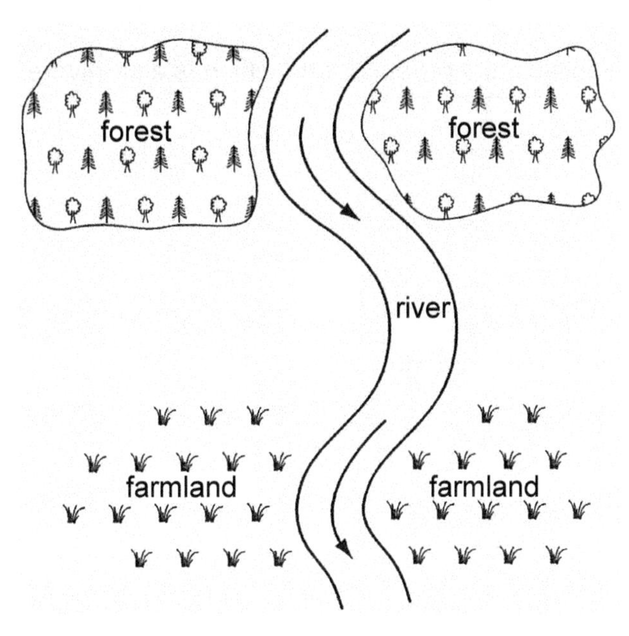

The forest is cut down.

What will be the most likely effect and how it is connected to the cutting down of the forest?

	likely effect	how it is connected to the cutting down of the forest
A	flooding of the farmland	water running off cleared area
B	gradual change to desert conditions	global warming
C	increase in number of water plants	more light falling on river
D	water logging of cleared area	trees no longer transpiring

( )

[N05/1/31]

23. The diagram shows nine organisms forming a food web.

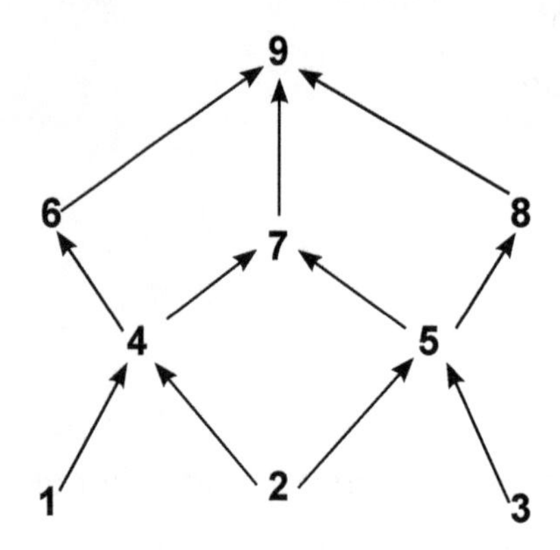

Which of the organisms is a producer and which is a carnivore?

	producer	carnivore
A	1	4
B	2	6
C	5	8
D	7	9

( )

[N05/1/32]

STRUCTURED QUESTIONS . . . .

1. The figure shows the energy flow in kilojoules (kJ) through a food chain.

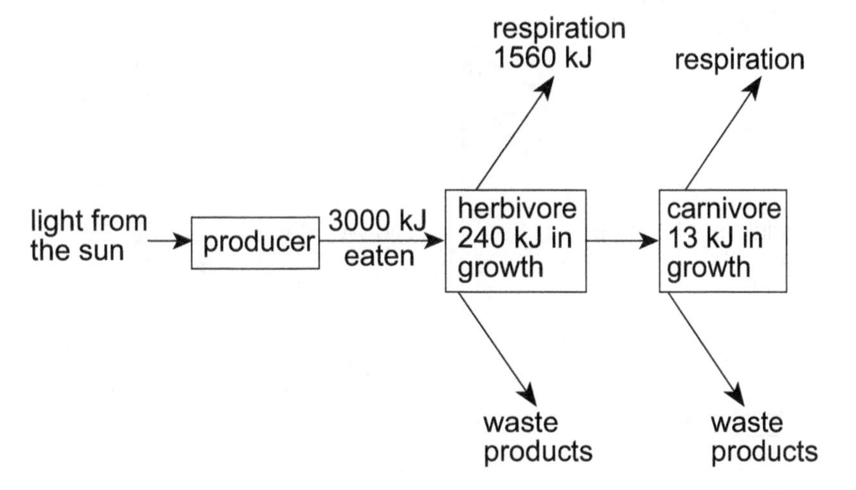

(a) (i) How many kJ are lost from the food chain in waste products from the herbivores?

....

....[1]

(ii) Calculate the percentage of the energy taken in by the herbivore that is used in growth.

....[2]

(b) State the importance of chlorophyll in the food chain.

....

....[1]

(c) Suggest why this food chain could not have another trophic level.

....

....[1]

[N08/2/8]

2. **Fig. (a)** shows part of a food web in a forest.

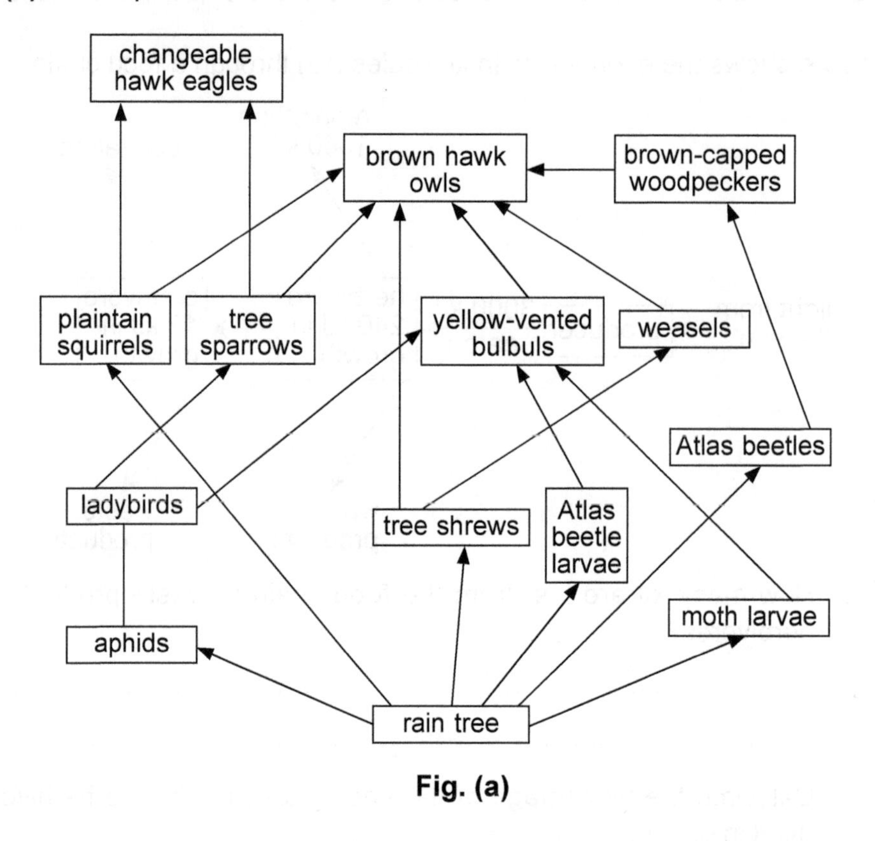

Fig. (a)

(a) Using an example from the food web, explain the term 'consumer'.

....

....[1]

(b) Caterpillars are herbivores and are preyed upon by tree sparrows and spiders.

Add this information to the food web. [2]

(c) During one particular year, most of the Atlas beetle larvae are killed by a disease. Explain how this could affect the brown-capped woodpecker population.

....

....

....

....

....[4]

(d) **Fig. (b)** shows one food chain from the food web.
The figures show the relative numbers of organisms at each level.

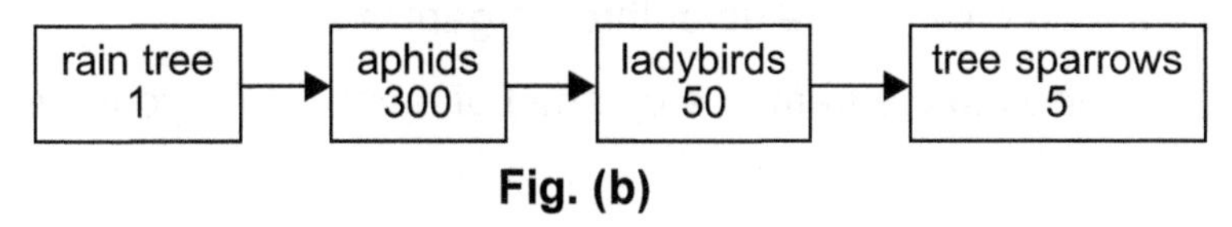

Fig. (b)

On the graph paper, complete the pyramid of numbers for this food chain.

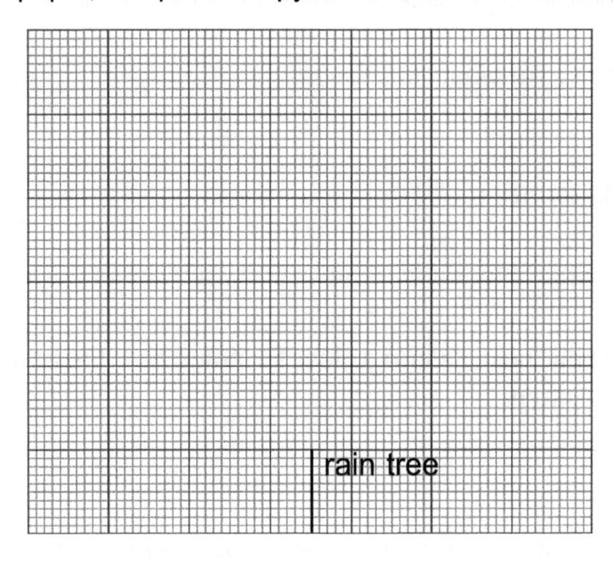

[2]

(e) List in sequence, four processes which occur to make available the carbohydrates in the body of the ladybirds for use by the tree sparrows in flying.

....

....

....[1]

[N07/2/1]

FREE RESPONSE QUESTIONS. . . . .

1. **(a)** Draw a food chain made up of three organisms.

Draw a **labelled** diagram of a pyramid of numbers for your food chain.

[3]

(b) Explain how energy losses occur along food chains. [4]

(c) Explain the term *bioaccumulation*. [3]

[N10/2/10 Or]

2. **(a)** Define the terms

(i) *producer*, [2]

(ii) *consumer*. [2]

(b) Describe how the carbon in a glucose molecule in the body of an animal is cycled in an ecosystem.

[5]

[N09/2/10 Or]

3. **(a)** Describe what is meant by the term *pyramid of numbers*. [3]

(b) **(i)** Describe how the organisms in a food chain or a food web form what is known as a pyramid of biomass.

(ii) Describe the part played by decomposers in a food web. [7]

[N05/2/8 Either]

4. Explain the effects that sewage can have on rivers and seas. [3]

[N05/2/8 Or (a)]

BIOLOGY

5094/01

Paper 1

October/November 2013

1 hour

1. The diagram shows a plant cell.

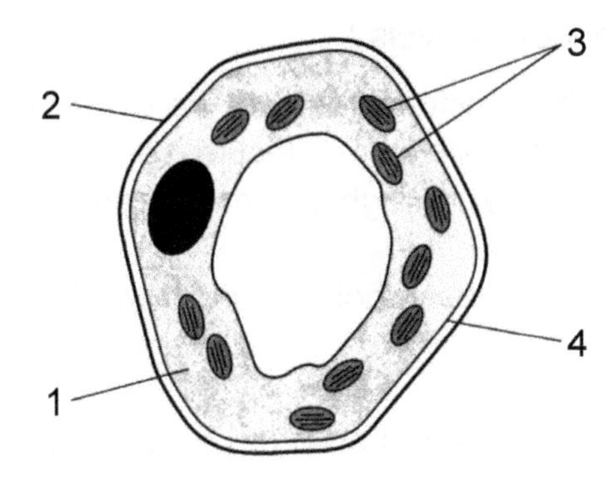

Which features are also found in an animal cell?

A 1 and 3
B 1 and 4
C 2 and 3
D 2 and 4

( )

2. Which row identifies the functions of the membrane systems and organelles listed?

	endoplasmic reticulum	Golgi body	mitochondrion	ribosome
A	energy release	collect and sort molecules	protein transport in cell	protein synthesis
B	protein synthesis	protein transport in cell	collect and sort molecules	energy release
C	protein transport in cell	collect and sort molecules	energy release	protein synthesis
D	protein transport in cell	energy release	protein synthesis	collect and sort molecules

( )

3. The diagrams show the apparatus used during an investigation.

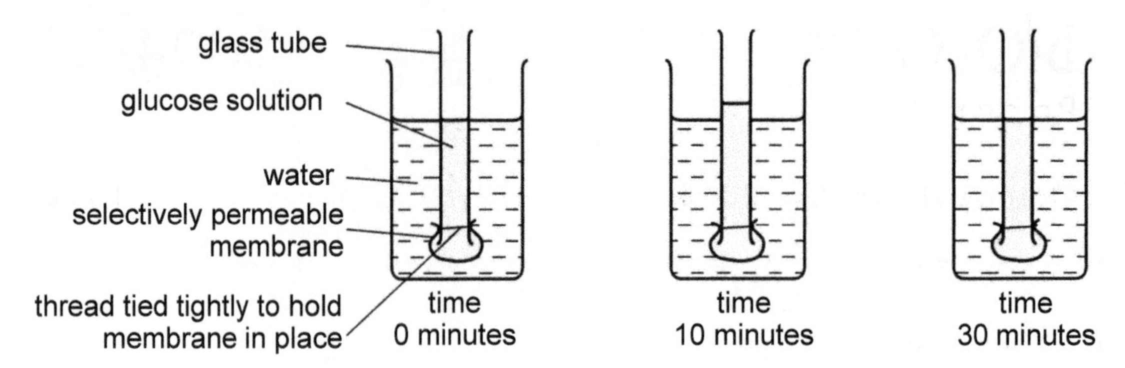

What explains the changes in levels?

A Glucose molecules have diffused against the concentration gradient.
B The membrane only allows water molecules to diffuse through it.
C Water molecules diffuse through the membrane more rapidly than glucose molecules.
D Water molecules have diffused both up and down the concentration gradient.

( )

4. Which chemical element forms part of all protein molecules?

A calcium
B iron
C magnesium
D nitrogen

( )

5. In the human body, large molecules are synthesised from small molecules.

Which row is correct for the small molecules required for synthesis of glycogen, lipids and proteins?

	glycogen	lipids	proteins
A	amino acids	glucose	glycerol and fatty acids
B	glycerol and amino acids	fatty acids	glucose
C	glycerol and fatty acids	glycerol and amino acids	fatty acids
D	glucose	glycerol and fatty acids	amino acids

( )

6. The diagrams represent the activity of an enzyme.

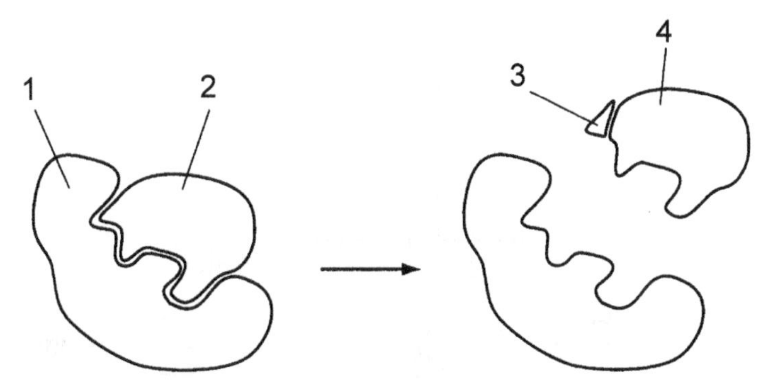

What are the labelled structures?

	'lock'	'key'	substrate	product
A	1	2	2	4
B	2	1	1	3
C	2	1	3	4
D	4	3	1	2

( )

7. Which row shows the conditions in the region of the alimentary canal where food is digested to produce **both** amino acids and fatty acids?

	pH	enzymes present
A	acid	amylase and lipase
B	acid	lipase and protease
C	alkaline	amylase and protease
D	alkaline	lipase and protease

( )

8. Which section of the diagram represents the effects of excessive alcohol consumption on the body?

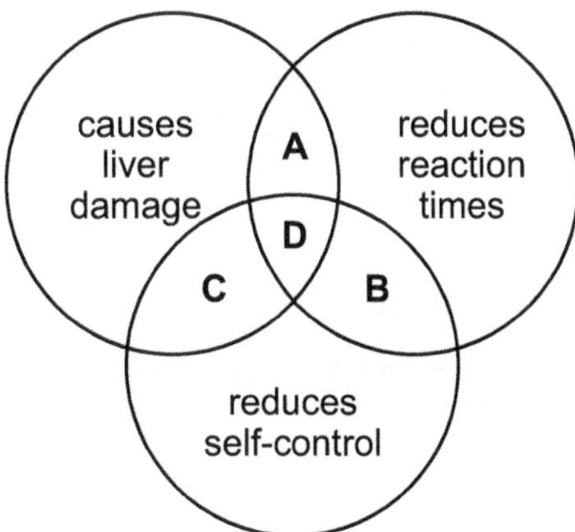

( )

9. The diagram shows an investigation using an aquatic plant.

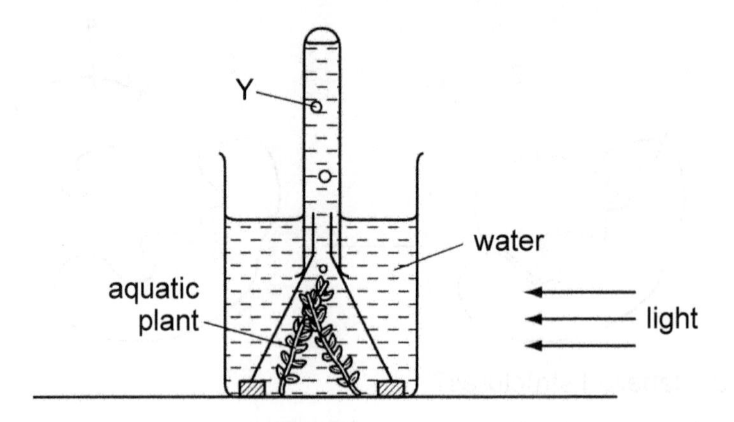

What is the main gas present at Y?

A air
B carbon dioxide
C nitrogen
D oxygen

( )

10. The graph shows the rate of photosynthesis at two different carbon dioxide concentrations and at varying light intensities at an optimum temperature of 25°C.

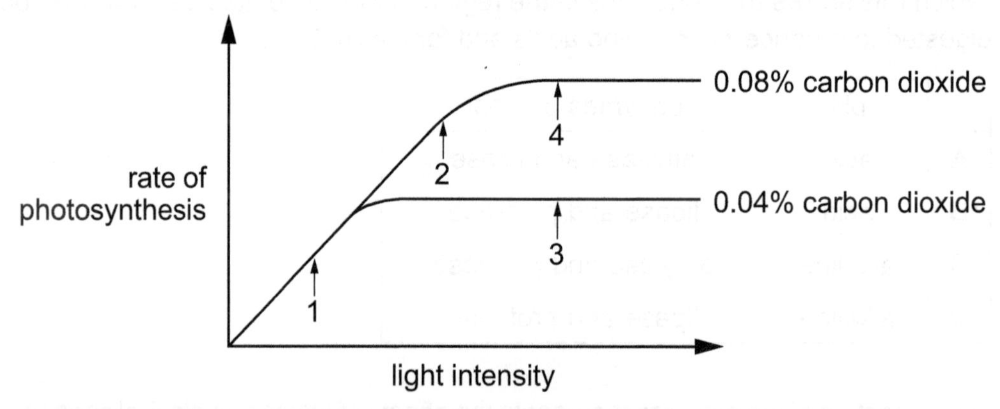

Which factors are limiting at the points indicated?

	light intensity	carbon dioxide concentration
A	1	2
B	3	4
C	1 and 2	3 and 4
D	2 and 3	1 and 4

( )

11. The diagram shows a section through a stem.

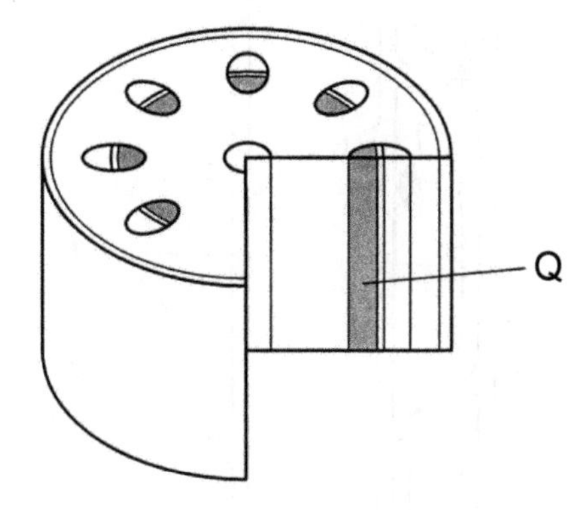

What is the main transport function of tissue Q?

	substance transported	carried from	carried to
A	sugar	roots	leaves
B	sugar	leaves	roots
C	water	roots	leaves
D	water	leaves	roots

( )

12. When young leaves are being formed on a plant, large quantities of mineral ions are needed.

Where and when is the movement of mineral ions in the plant the greatest?

A companion cells on a hot sunny day
B root hair cells on a cool cloudy day
C sieve tube elements during a warm night
D xylem vessels on a warm sunny day ( )

13. The diagram shows a section through part of a stem of a plant.

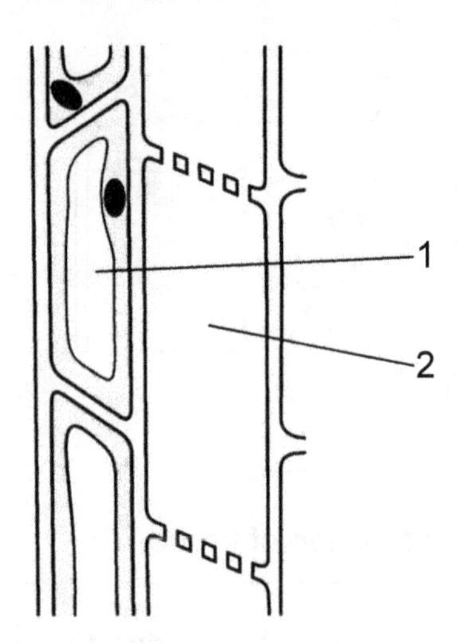

Which row gives the correct names of the numbered structures?

	structure 1	structure 2
A	companion cell	sieve tube element
B	companion cell	xylem vessel
C	sieve tube element	companion cell
D	xylem vessel	sieve tube element

( )

14. The graph shows changes in blood pressure as blood flows through the blood vessels of the human circulatory system.

Which blood vessel, **A**, **B**, **C** or **D**, contain the most elastic tissue in their walls?

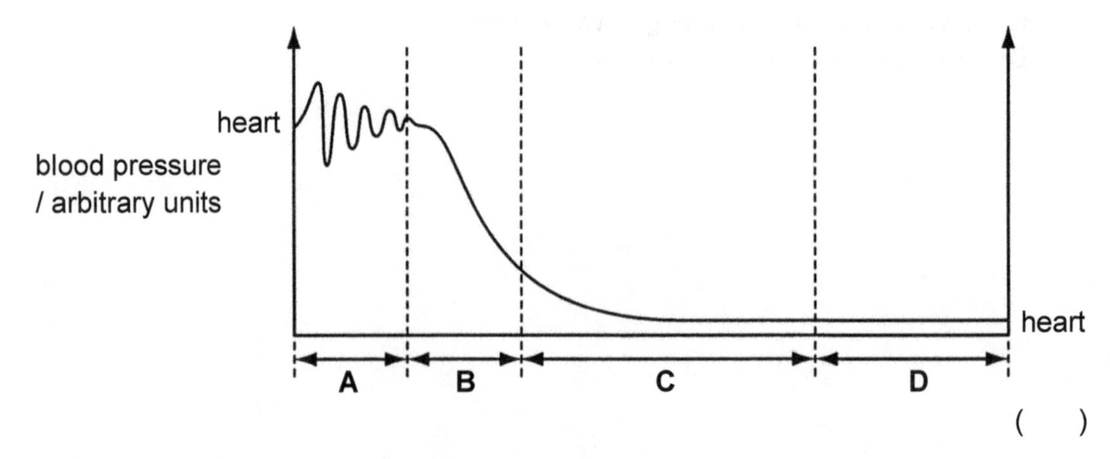

( )

15. The diagram shows part of the human circulatory system.

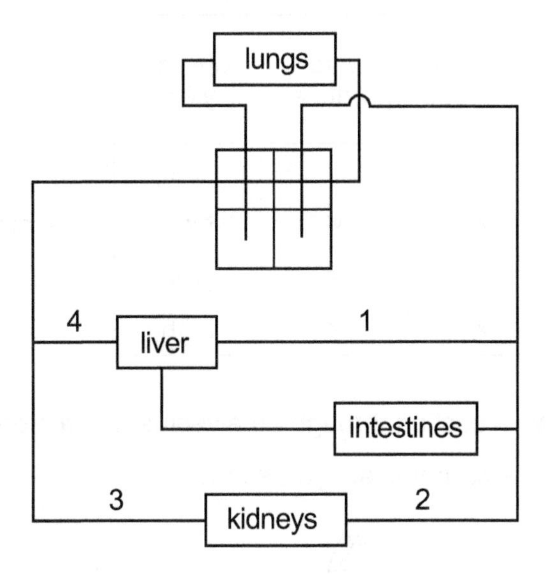

Which blood vessels contain blood with the highest and lowest concentrations of urea?

	highest concentration of urea	lowest concentration of urea
A	1	4
B	2	1
C	3	2
D	4	3

( )

16. The diagram shows some tissue cells next to a capillary in the body of a human male.

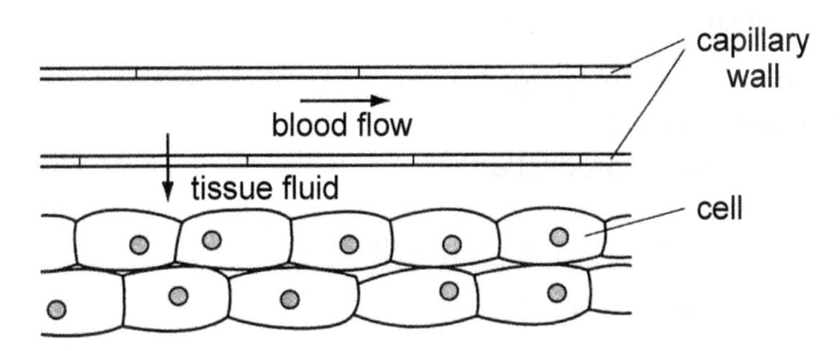

What is present in the tissue fluid formed from the plasma?

A ADH and haemoglobin
B glucose and oxygen
C glucose and platelets
D progesterone and water

( )

17. Which row is correct for anaerobic respiration of glucose in human muscle?

	carbon dioxide produced	lactic acid produced	water produced	energy yield
A	✓	✓	×	low
B	✓	×	✓	high
C	×	✓	×	low
D	×	×	✓	high

key
✓ = produced
× = not produced

( )

18. The diagram shows a section through an alveolus and a blood capillary.

Where is the enzyme carbonic anhydrase found?

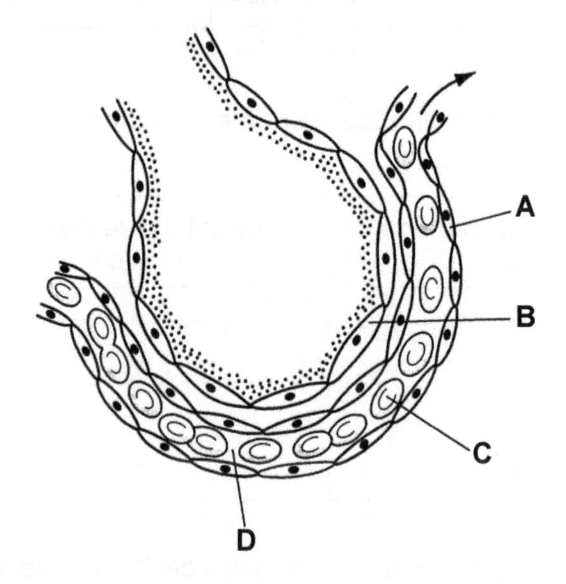

( )

19. Some effects of smoking are listed.

1 causes uncontrollable cell division
2 increases heart rate
3 increases mucus production
4 is addictive
5 reduces the amount of oxygen in the blood

Which effects are caused by nicotine?

A 1 and 3
B 2 and 4
C 1, 2 and 5
D 3, 4 and 5

( )

20. Which statement about excretory materials is correct?

A All nitrogenous compounds must be excreted.
B They always contain the element nitrogen.
C They are always present in excess in the diet.
D They are produced by the cells in the body. ( )

21. Two samples of fluids were removed from different parts of a kidney tubule and analysed. The results, in arbitrary units, are shown in the table.

chemical	glomerular filtrate	second sample
urea	10	8
sodium ions	10	1
water	100	5
glucose	5	0

From which position was the second sample taken?

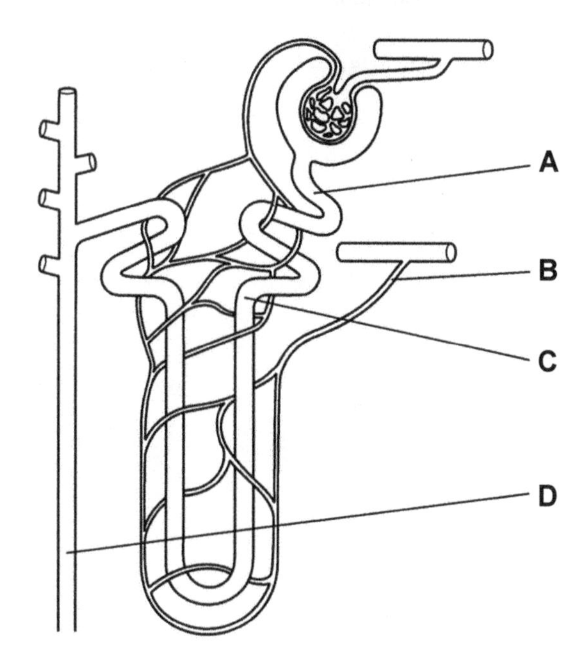

( )

22. The diagram shows a section through the skin.

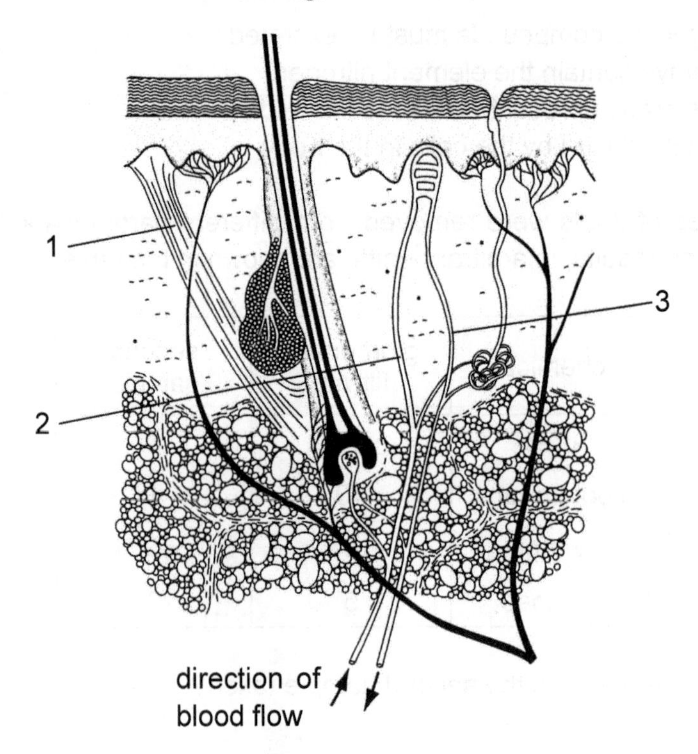

Which structure(s) contain(s) muscle that contracts when the body is too cold?

A 1 only
B 1 and 2 only
C 1 and 3 only
D 2 only ( )

23. The diagram represents the movement of water between the environment and some living organisms.

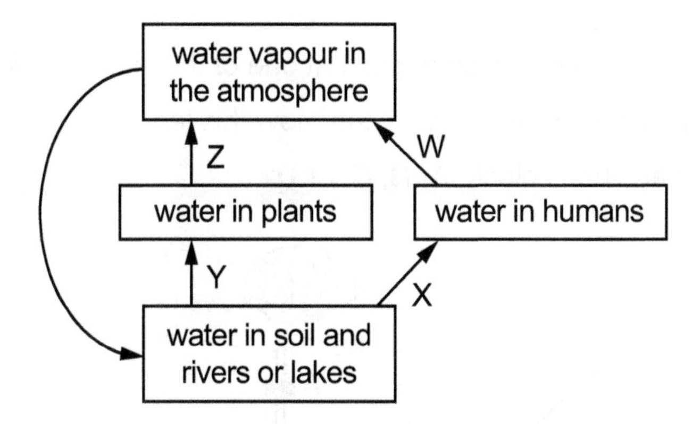

Where are the processes of transpiration and respiration involved in the movement of water?

	respiration	transpiration
A	W	Z
B	X	Y
C	Y	W
D	Z	X

( )

24. What results in the release of insulin?

A a decrease in the level of glucagon in the blood
B an increase in the level of adrenaline in the blood
C an increase in the level of glucose in the blood
D an increase in the number of nerve impulses passing to the pancreas ( )

25. This diagram of the nervous system shows four places, **A**, **B**, **C**, and **D**, where a local anaesthetic block can be applied. The block prevents nerve impulses travelling along neurones.

A man had an anaesthetic block applied at one of the sites shown, **A**, **B**, **C**, or **D**.

He can feel a pinprick on his leg and can move his leg.

Where is his anaesthetic block, **A**, **B**, **C**, or **D**?

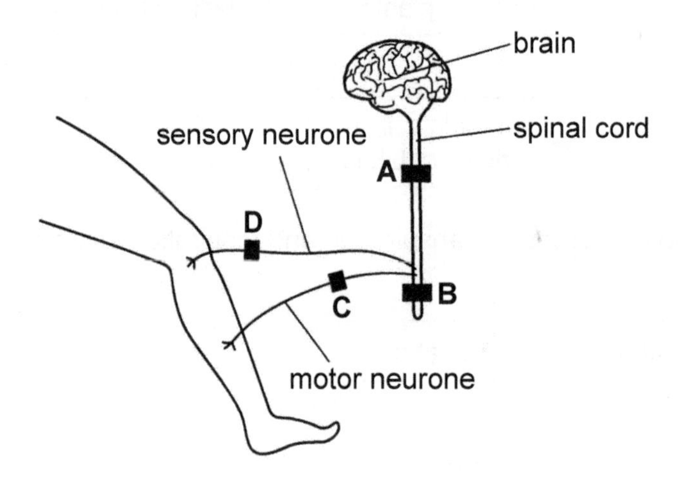

( )

26. The diagrams show two sections through the eye of the same person.

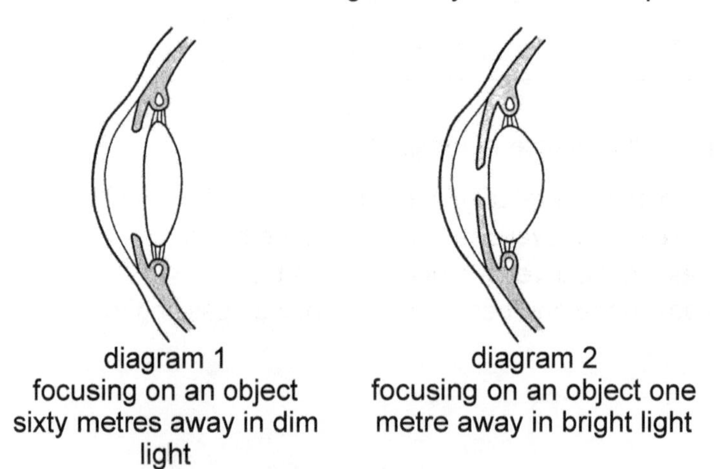

diagram 1
focusing on an object sixty metres away in dim light

diagram 2
focusing on an object one metre away in bright light

What happens to achieve the changes from the eye in diagram 1 to the eye in diagram 2 under the different conditions?

	ciliary muscles	iris radial muscles	iris circular muscles
A	contract	contract	relax
B	contract	relax	contract
C	relax	contract	relax
D	relax	relax	contract

( )

27. The diagram shows the structure of a carpel, pollen grain and pollen tube just before fertilisation.

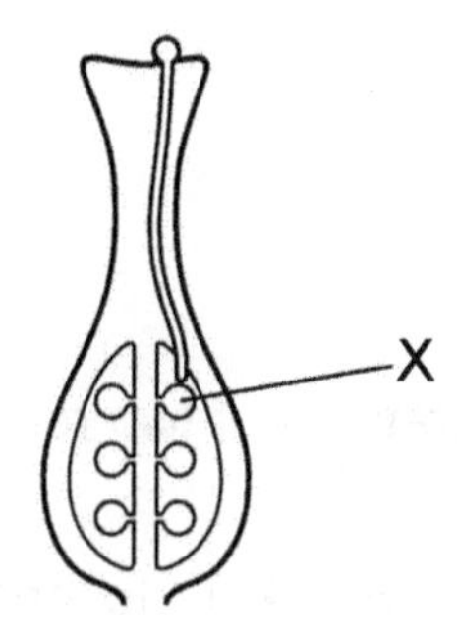

What is the name of the structure labelled X?

A ovary
B ovule
C ovum
D seed ( )

28. The diagram shows a section through a flower.

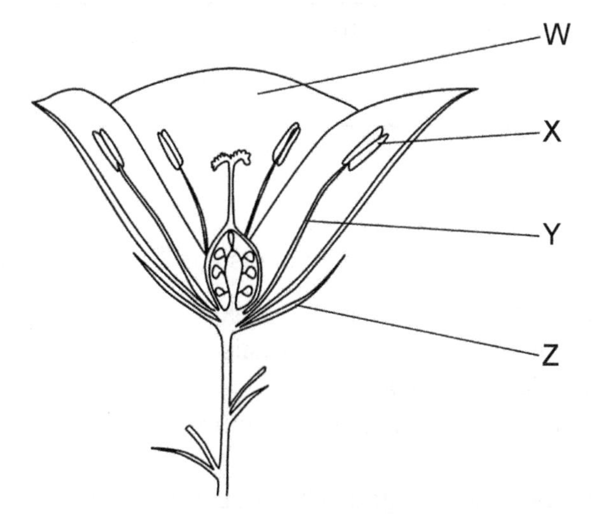

What are the functions of the parts labelled W, X, Y and Z?

	W	X	Y	Z
A	attracts insects	makes pollen	supports anther	protects other flower parts
B	makes pollen	supports anther	protects other flower parts	attracts insects
C	protects other flower parts	attracts insects	makes pollen	supports anther
D	supports anther	protects other flower parts	attracts insects	makes pollen

( )

29. Which part of the male reproductive system helps to maintain a low temperature in the testes?

A prostate gland
B scrotum
C sperm ducts
D urethra ( )

30. The graph shows the concentrations of two reproductive hormones in the blood of an adult female.

During which period is she most likely to become pregnant if she has sexual intercourse?

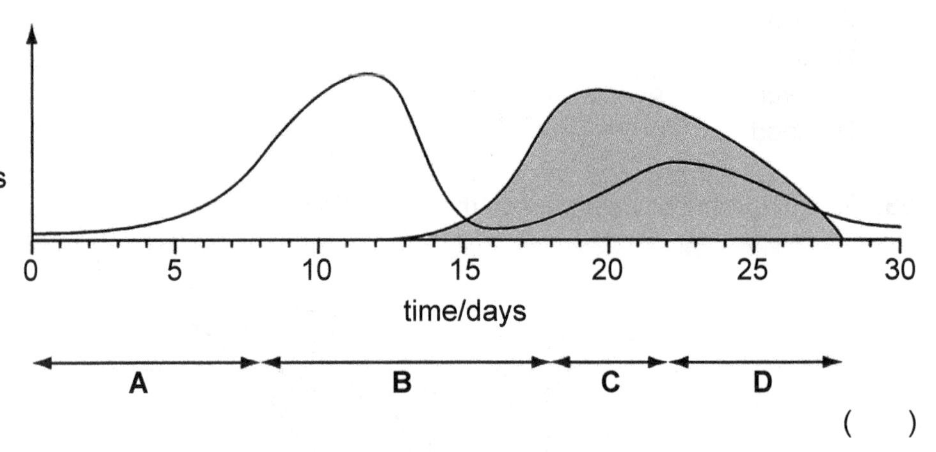

( )

31. When does a cell division occur that involves a reduction in chromosome numbers?

A during the formation of new cells in the skin
B during the formation of sperm in the testis
C in the haploid cells in the ovary
D in the zygote immediately after fertilisation ( )

32. Which statement about homologous chromosomes is correct?

A They contain identical DNA.
B They contain the same genes in the same position.
C They have identical alleles.
D They replicate during meiosis only. ( )

33. A cell containing three pairs of chromosomes divides by meiosis.

Which diagram shows one of the daughter cells after telophase II?

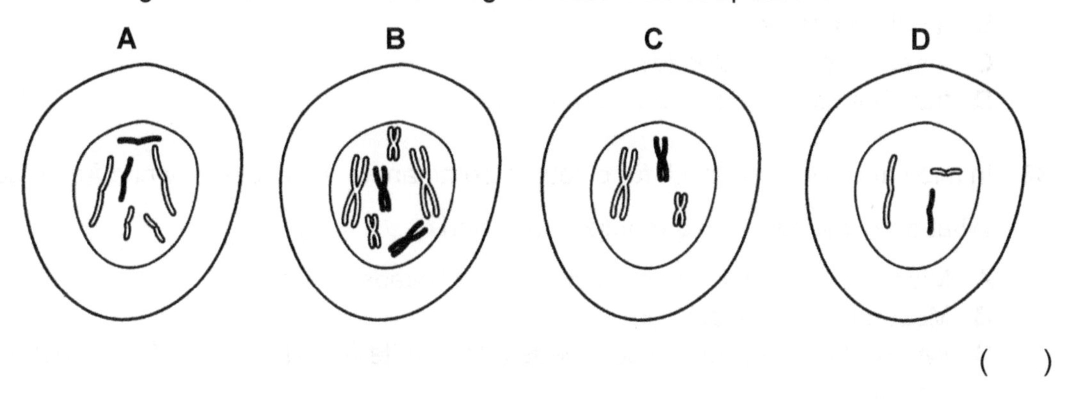

( )

34. The diagram represents a nucleotide consisting of a sugar, a phosphate group and an organic base.

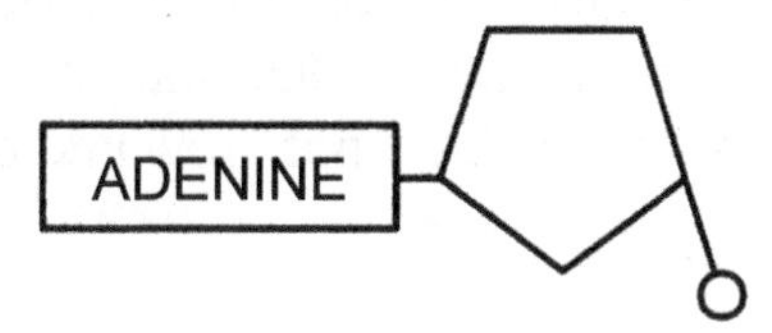

In a DNA molecule, which nucleotide will pair with the above nucleotide?

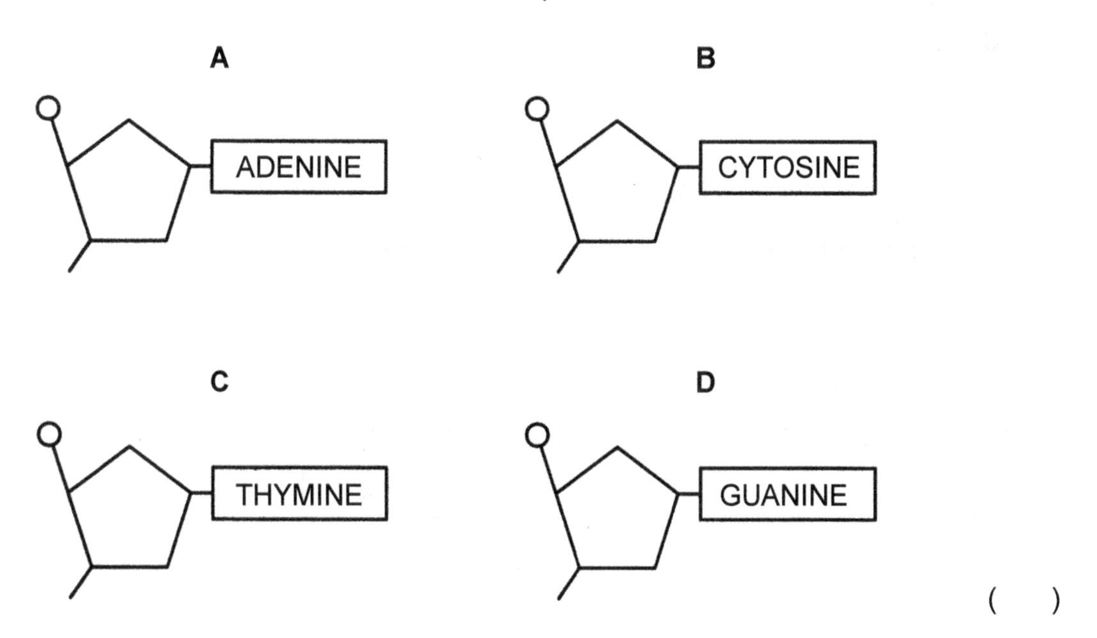

( )

35. Which statement describes a gene?

A a number of DNA molecules
B a pair of alleles
C a sequence of nucleotides
D the chain of alleles on a chromosome ( )

36. In mice, the allele for black fur colour is dominant to the allele for white fur colour.

What does this mean in a mouse population?

A Mice with black fur are more successful breeders.
B Most mice have black fur.
C When a black-furred mouse breeds with a white-furred one, the offspring will have grey fur.
D White-furred mice are only born to two white-furred parents. ( )

37. A species of fish has the strength of its bones controlled by alleles B and b. Fish with the genotype BB have very strong bones, fish with the genotype Bb have thin, weak bones and fish with the genotype bb do not grow into an embryo.

Two heterozygous fish mate.

Which proportion of their offspring will be heterozygous?

A 25%
B 50%
C 67%
D 75% ( )

38. Which fertilisation would result in a male child with Down syndrome?

	chromosomes in ovum	fertilised by	chromosomes in sperm
A	22 + 1X		22 + 1Y
B	22 + 1X		23 + 1Y
C	23 + 1Y		22 + 1X
D	23 + 1Y		23 + 1X

( )

39. The diagram shows part of a food web in a freshwater pond.

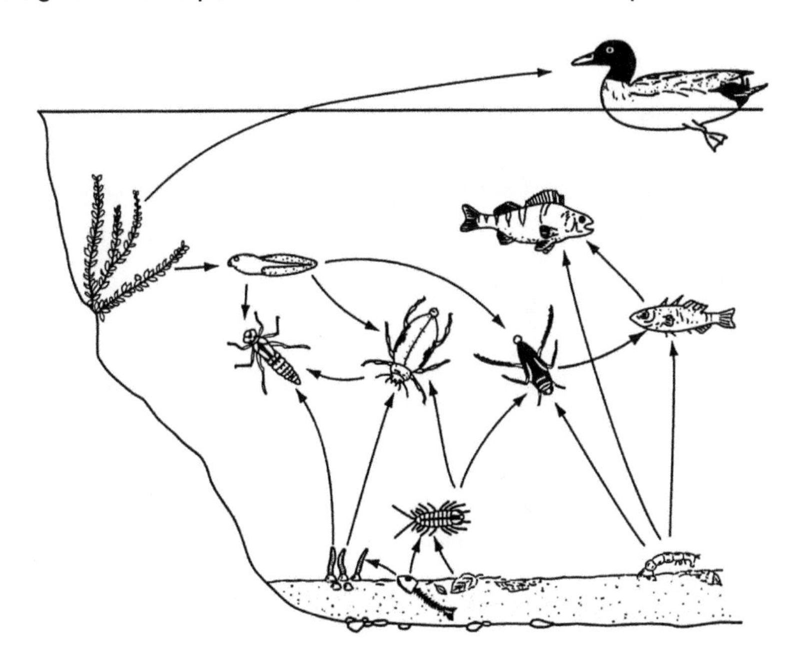

In this web, which is **not** correct?

A One trophic level is herbivorous.
B Two trophic levels are producers.
C Three trophic levels are carnivorous.
D Four trophic levels are consumers. ( )

40. The diagram shows a pyramid of numbers in an ecosystem on land.

Which organisms are smallest in body size?

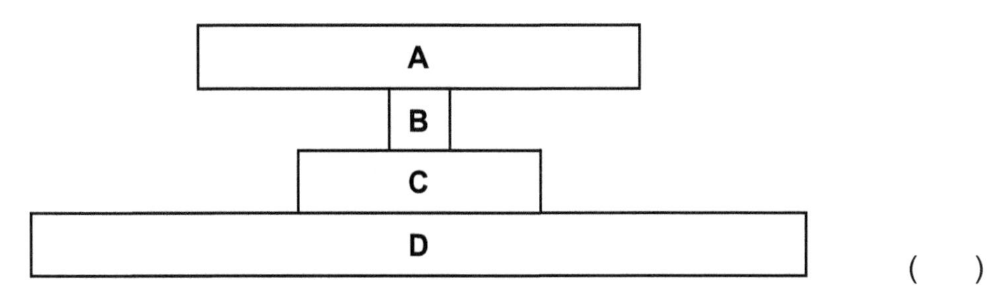

( )

BIOLOGY

5094/02

Paper 2

October/November 2013

1 hour 45 minutes

Section A

Answer **all** questions.

Write your answers in the spaces provided.

1. Fig. 1.1 shows a vertical section of the human heart.

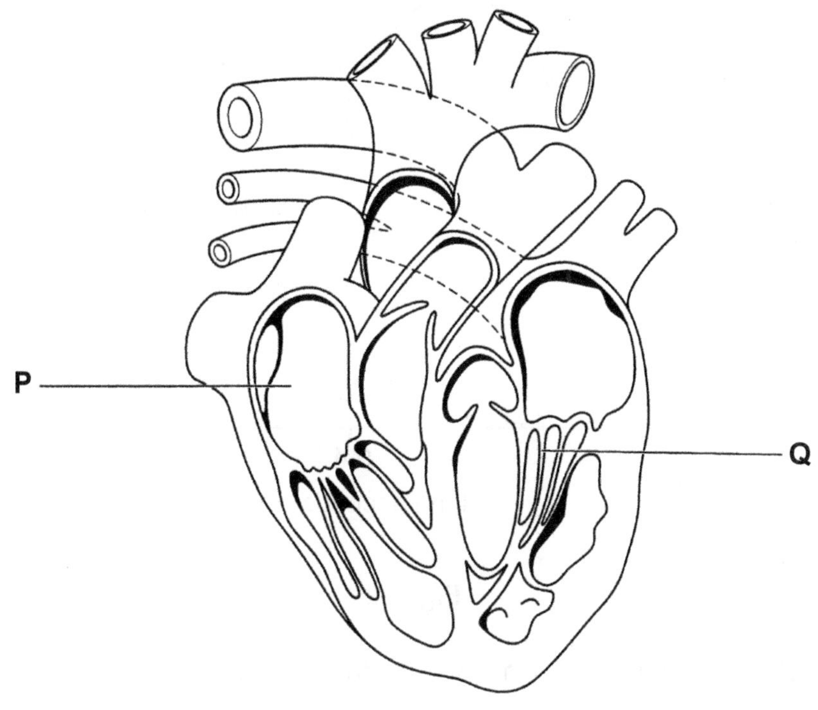

Fig. 1.1

(a) (i) Name the parts labelled **P** and **Q**.

P ....

Q ....

[2]

(ii) Use a label line and the letter **X** to label the aorta on Fig. 1.1. [1]

(b) Describe how blood from the lungs is forced through the heart into the aorta.

....

....

....

....

....

....

....

.... [4]

(c) Fig. 1.2 shows the changes in pressure in the left side of the heart during the cardiac cycle.

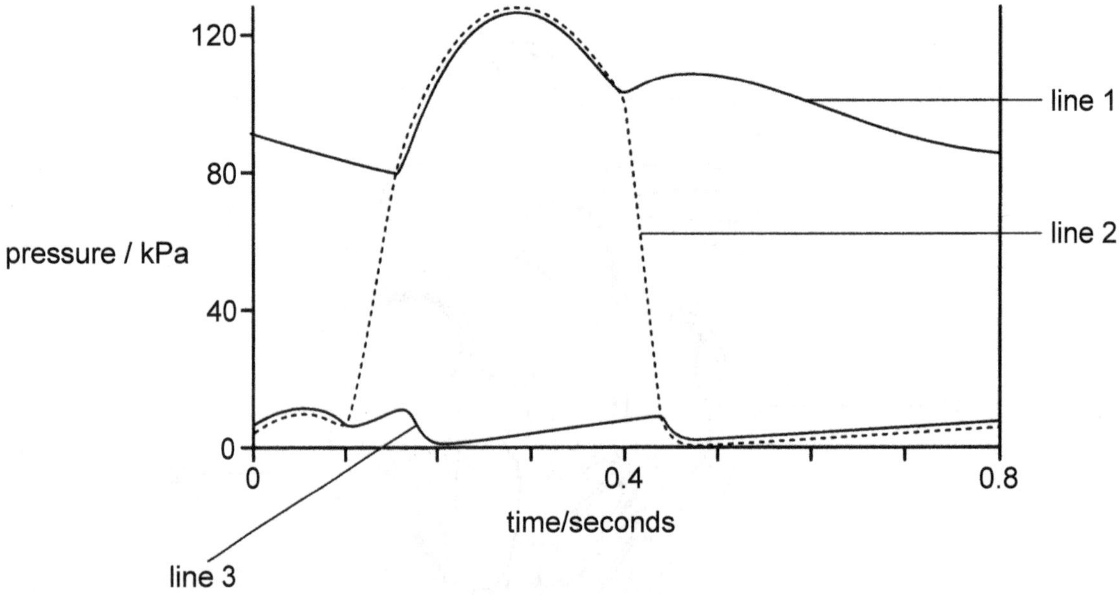

Fig. 1.2

(i) Using the information in Fig. 1.2, state how long ventricular systole lasts.

....[1]

(ii) State the number of the line which represents the pressure in the left ventricle.

....[1]

[Total: 9]

2. **(a)** Define the term *translocation*.

....

....

....

.... [2]

(b) Fig. 2.1 shows a transverse section of part of the stem of *Helianthus annuus*.

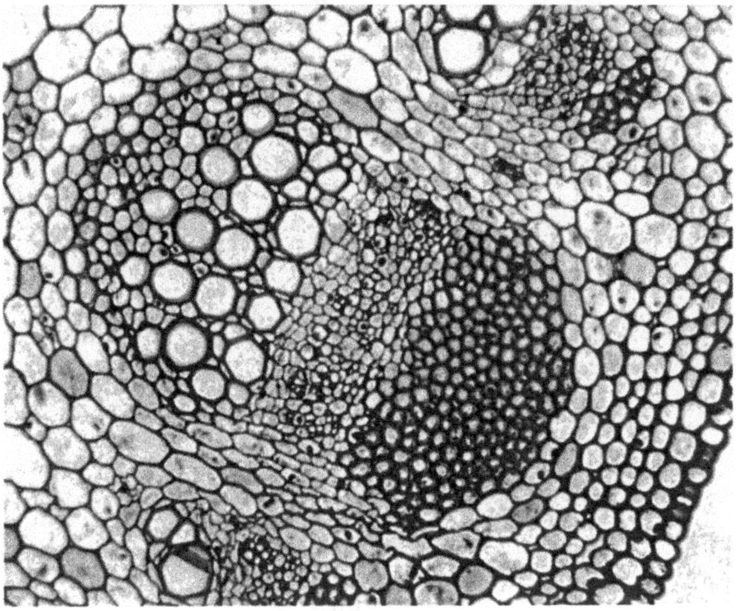

Fig. 2.1

Use label lines and labels to identify the position on Fig. 2.1 of:

(i) a xylem vessel

(ii) the phloem. [2]

(c) Some weedkillers stop the plant from photosynthesising.
These are often applied to the soil where the weeds are growing.
Explain how the weedkiller reaches its site of action in the leaves.

....

....

....

....

....

....

....[4]

[Total: 8]

3. Fig. 3.1 shows a kidney tubule.

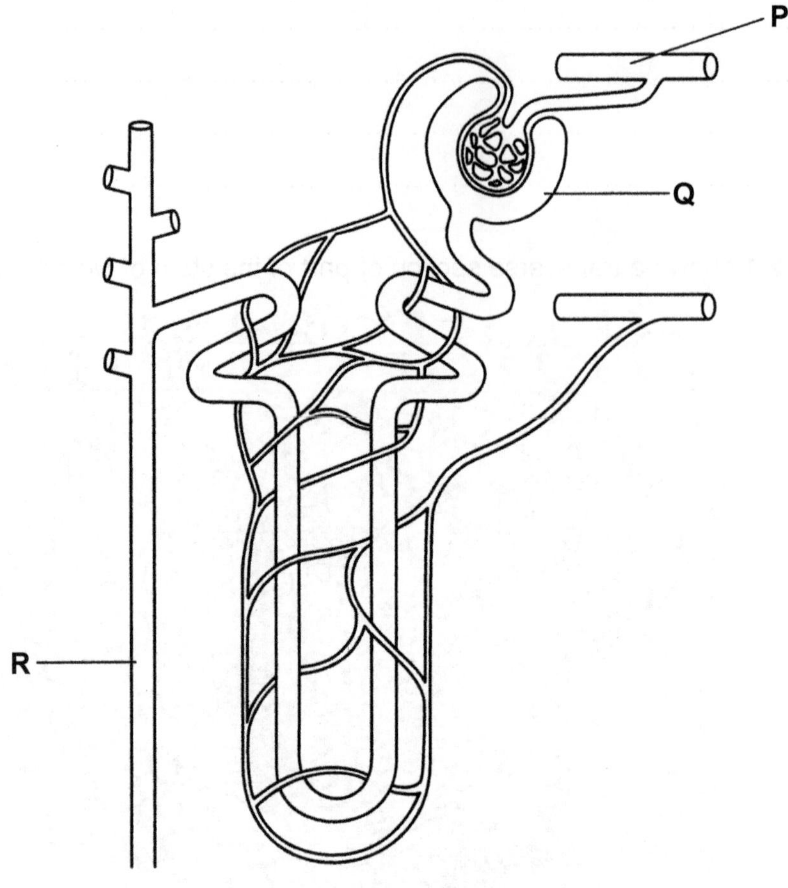

Fig. 3.1

(a) Name the structure labelled **P** on Fig. 3.1.

....[1]

(b) Table 3.1 shows some of the substances present at each of the regions **P**, **Q** and **R**.

Table 3.1

region	protein content g / 100 cm^3	glucose content g / 100 cm^3
P	8.0	0.1
Q		
R	0.0	0.0

(i) Complete the table to show the protein content and the glucose content at region **Q**. [2]

Explain the differences in:

(ii) the protein content between regions **P** and **R**.

....

....[2]

(iii) the glucose content between regions **Q** and **R**.

....

....[1]

(c) Explain how anti-diuretic hormone (ADH) affects the composition of the liquid passing through region **R**.

....

....

....

....

....

....[3]

[Total: 9]

4. **(a)** State what is meant by the term *activation energy*.

....

....[1]

(b) Fig. 4.1 shows a section of photographic film.
The top layer is made up of silver particles embedded in a layer of gelatine which is a type of protein.

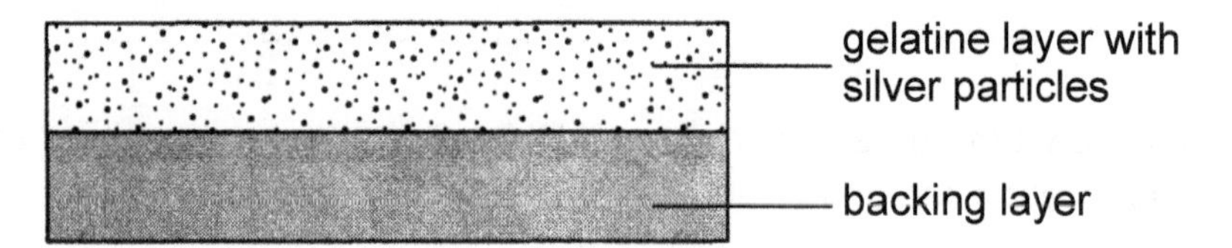

Fig. 4.1

In an investigation a 20 mm length of photographic film was placed into each of three boiling tubes.

- The film was immersed in 20 cm^3 water.
- 1 cm^3 of liquid at different pH values was added to the boiling tubes.
- 1 cm^3 of protease solution was added to each boiling tube.
- Each boiling tube was shaken gently to mix the contents.
- Each boiling tube was kept at 37°C for 1 hour.

Fig. 4.2 shows the apparatus and the results of the investigation.

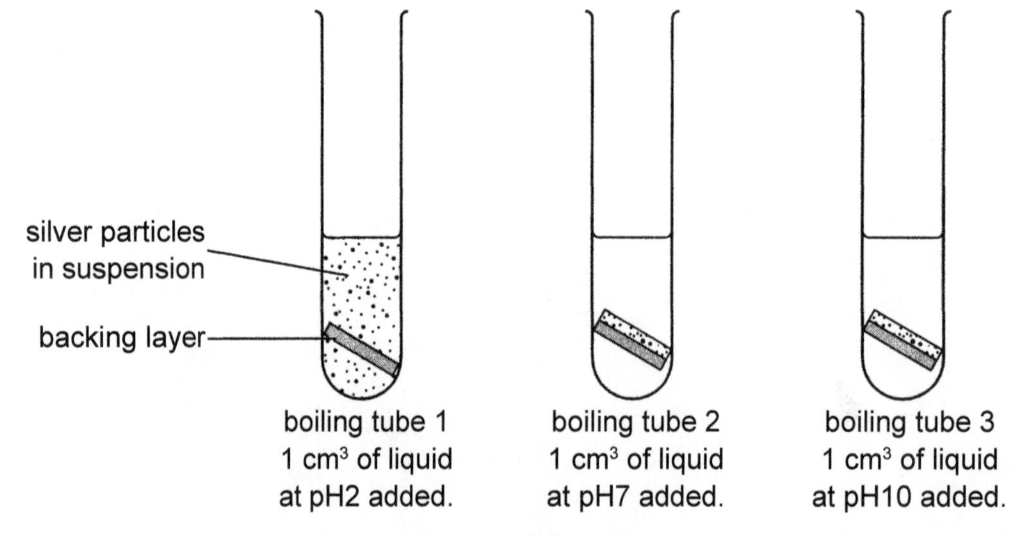

Fig. 4.2

(i) Explain the results in boiling tubes 1, 2 and 3.

boiling tube 1 ....

....

....

....

....

....[2]

boiling tube 2 and boiling tube 3 ....

....

....

....

....

....[2]

(ii) State two factors kept constant during the investigation.

1. ....

2. ....

[2]

[Total: 7]

5. Fig. 5.1 shows a food web in a river.

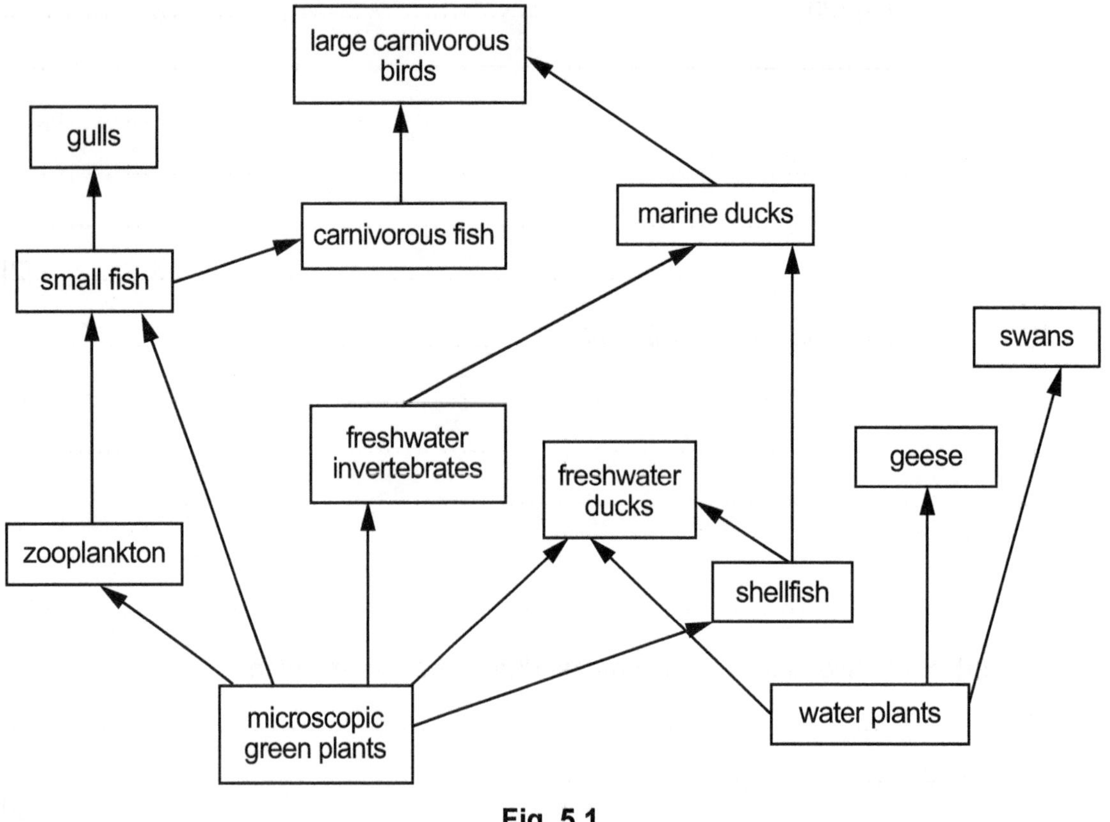

Fig. 5.1

(a) From Fig. 5.1 select a food chain of five organisms.
Draw the food chain in the space below. [2]

(b) (i) Name the producers in the food web in Fig. 5.1

....[1]

(ii) With reference to Fig. 5.1, explain the term producer.

....

....

....

....

....[3]

[Total: 6]

6. Fig. 6.1 shows the inheritance of eye colour which is controlled by a single pair of alleles.

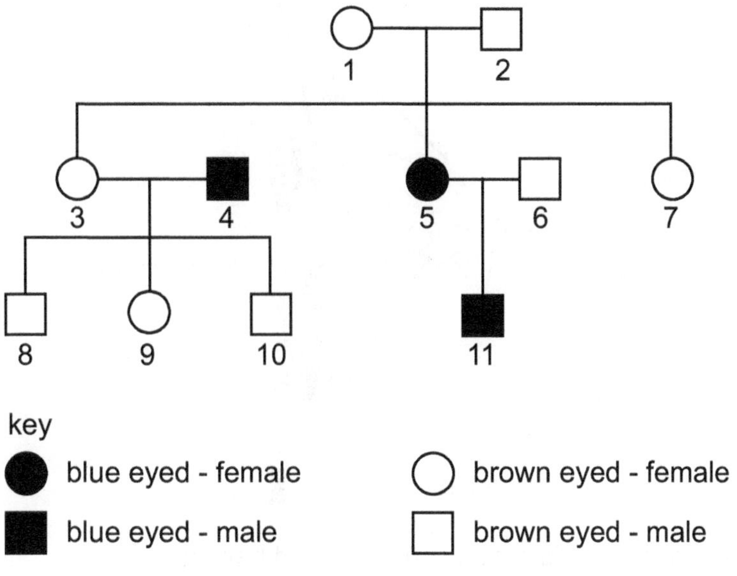

Fig. 6.1

(a) Using **B** to represent the allele for brown eye colour and **b** to represent the allele for blue eye colour, state with reasons the genotype of individuals 1 and 10.

individual 1 ....

....

....

....[2]

individual 10 ....

....

....

....[2]

(b) Using the numbers 1 to 11 for the individuals shown in Fig. 6.1, identify the males that are heterozygous for eye colour.

....[1]

(c) State the type of variation shown by eye colour.

....[1]

[Total: 6]

7. Fig. 7.1 shows a section of a mammary gland.

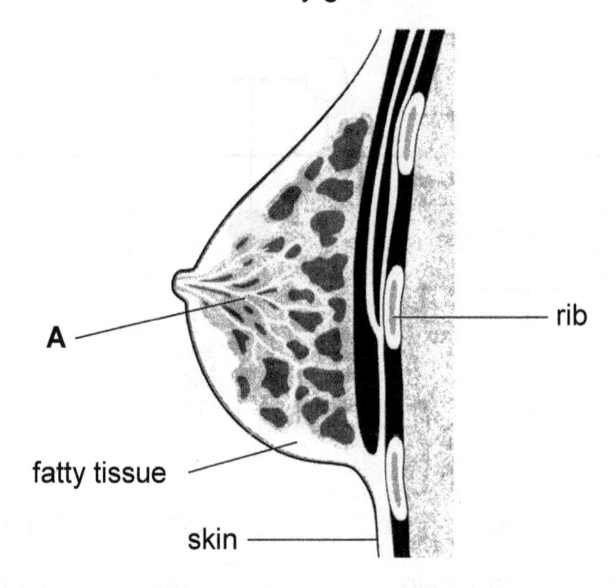

(a) (i) Mammary glands produce milk to feed babies.
The production and secretion of milk is stimulated by the hormone prolactin.
Define the term *hormone*.

....

....

....[3]

(ii) Suggest a function for the structure labelled **A**.

....

....[1]

(b) Table 7.1 shows the composition of human and cow's milk.

Table 7.1

substance	human milk %	cow's milk %
water	88.4	87.3
sugar	6.9	4.6
protein	1.2	3.5
fat	3.3	4.0
mineral salts	0.2	0.6

Using the information in Table 7.1, state the substances that must be **added to** cow's milk to make its composition closer to that of human milk.

....[1]

[Total: 5]

Section B

Answer **three** questions.

Question 10 is in the form of an **Either/Or question**.
Only one part should be answered.

8. Table 8.1 shows the relationship between mean body length and mean rate of heat production in some animals.

Table 8.1

mean body length / mm	200	400	600	800	1000	1200	1400
mean rate of heat production / kJ per kg per hour	27.5	19.5	13.0	8.6	6.0	4.5	3.6

(a) Plot these data on the graph paper. [4]

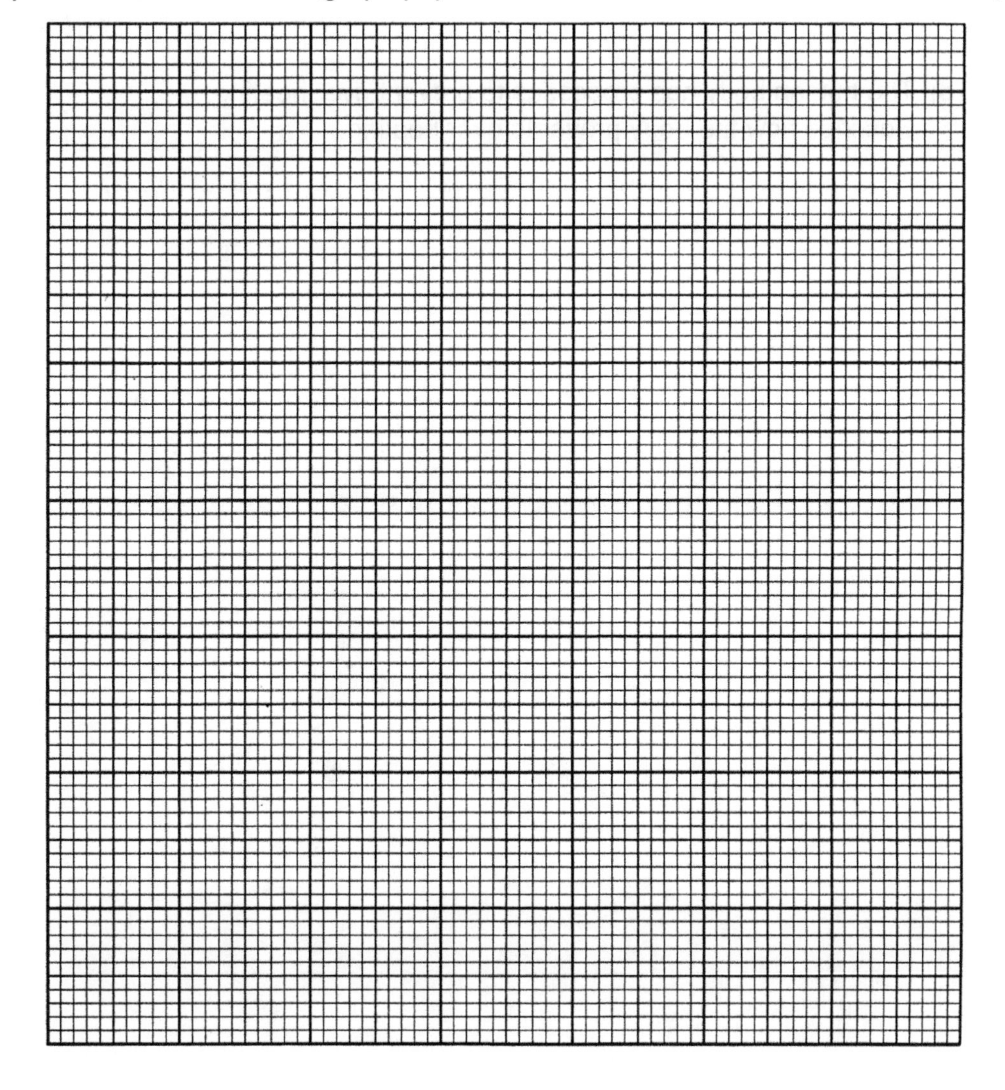

(b) (i) Use your graph to estimate the mean body length of an animal with a mean rate of heat production of 10.5kJ per kg per hour.

....[1]

(ii) An animal has a mean body length of 550 mm.
Use your graph to estimate the mean rate of heat production in this animal.

....[1]

(iii) State the relationship between mean body length and mean rate of heat production.

....

....[1]

(c) Name the chemical process which produces heat in the body.

....[1]

(d) The human body has processes to prevent over heating.
Describe how the body prevents over heating.

....

....

....

....

....

....

....

....

....

....[3]

[Total: 11]

9. **(a)** Describe what is meant by a *transgenic organism.*

....

....

....

....

....

....

....

....

....

.... [3]

(b) Suggest three reasons for the development of transgenic organisms.

1. ....

....

....

2. ....

....

....

3. ....

....

....[3]

(c) The large-scale production of transgenic bacteria is carried out in large vessels called fermenters.

Suggest three factors which should be kept constant in a fermenter.

1. ....

....

....

2. ....

....

....

3. ....

....

....[3]

[Total: 9]

10. Either

(a) Distinguish between sexual and asexual reproduction.

....

....

....

....

....

....

....

....[4]

(b) Define the term *pollination*.

....

....

....

....

....[2]

(c) State two differences between wind-pollinated and insect-pollinated flowers. Give a reason for each difference.

1. ....

....

....

2. ....

....

....[4]

[Total: 10]

10. Or

(a) Describe the process of *transpiration*.

....

....

....

....

....

....

....

....[4]

(b) Explain how each of the following factors affect transpiration:

(i) high humidity of the air

....

....

....

....

....

....[3]

(ii) an increase in temperature of the air.

....

....

....

....

....

....[3]

[Total: 10]

BIOLOGY 5158/01

Paper 1

October/November 2014 1 hour

1. The diagram shows an animal cell.

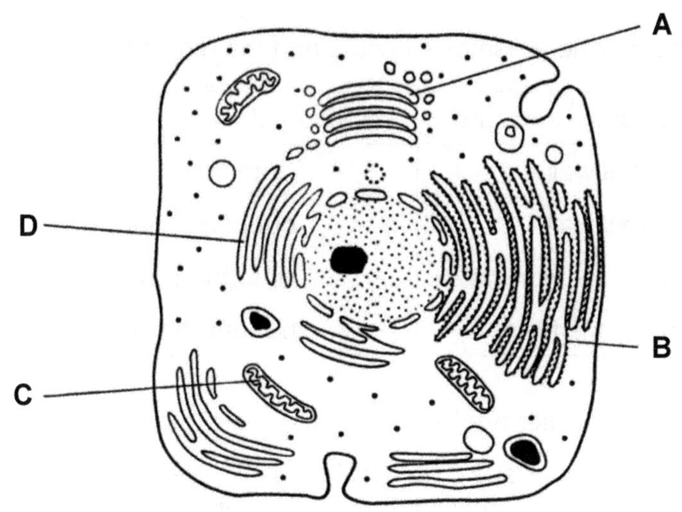

Which label links an organelle to its function?

organelle	function
A	packaging and processing of proteins
B	synthesis of fats
C	control of polypeptide synthesis
D	release of energy

( )

2. Which mature structure contains a nucleus?

A red blood cell
B root hair cell
C sieve tube element
D xylem vessel

( )

3. The diagram shows a plant cell in a 5% glucose solution.
The concentration of the solution in the vacuole is equivalent to a 10% glucose solution.

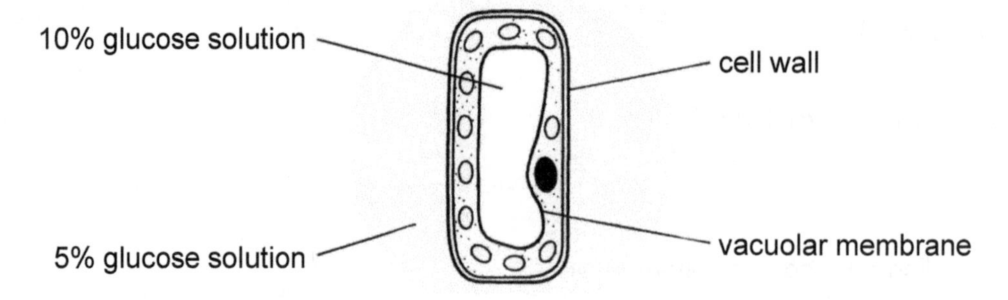

Which row states where osmosis occurs and the direction of water movement?

	where osmosis occurs	direction of water movement
A	cell wall	into the cell
B	cell wall	out of the cell
C	vacuolar membrane	into the cell
D	vacuolar membrane	out of the cell

( )

4. Three properties of water are listed.

1 Water cools a surface from which it evaporates.
2 Water is used as a solvent for many chemicals.
3 Water is involved in many metabolic reactions.

Which of these properties make water suitable to use in a blood transport system?

A 1 and 2 **B** 1 and 3 **C** 2 only **D** 3 only ( )

5. Which process can be carried out **without** the use of enzymes?

A conversion of starch to maltose in the mouth
B digestion of protein in the alimentary canal
C production of sugar in green leaves
D uptake of water by plant roots ( )

6. In which process are water molecules broken down and converted into other chemicals?

A aerobic respiration of glucose
B formation of carbohydrates in chloroplasts
C formation of lactic acid in muscles
D metabolism of amino acids ( )

7. The diagram shows the human alimentary canal with labels for the functions of some of its parts.

Which label is correct?

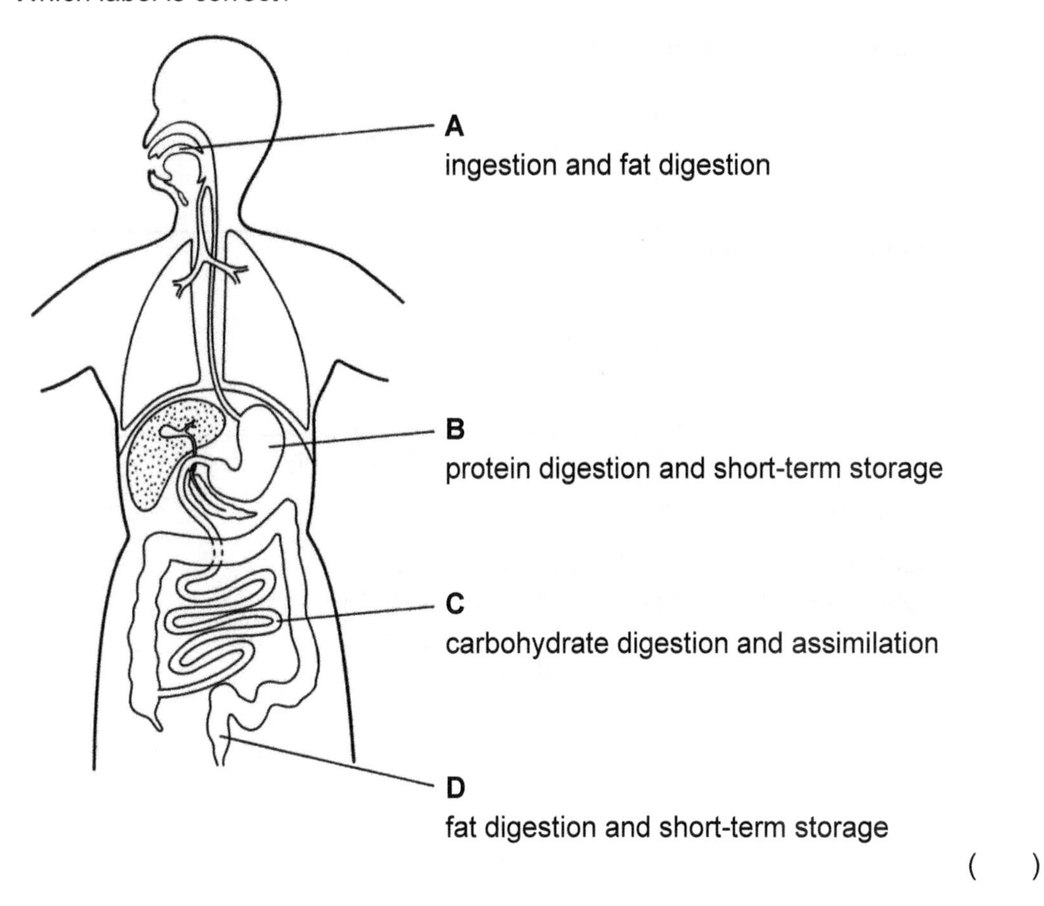

( )

8. Some processes that occur in the body are listed.

1 breakdown of red blood cells
2 breakdown of starch
3 formation of urine
4 storage of glycogen

Which processes occur in the liver?

A 1 and 2 **B** 1 and 4 **C** 2 and 3 **D** 3 and 4 ( )

9. An investigation was carried out on the effect of light intensity on the rate of photosynthesis.
Throughout the experiment, all other factors affecting photosynthesis were kept constant.
The results are shown on the graph.

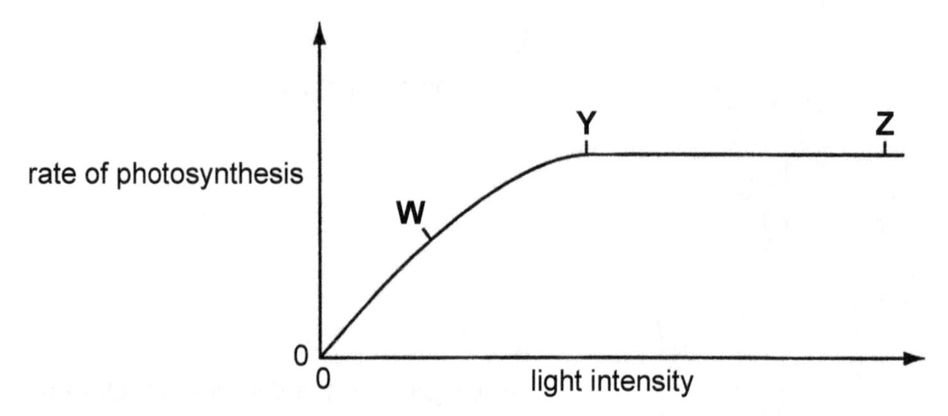

What do these results show?

A Light does not limit the rate of photosynthesis.
B Light limits the rate of photosynthesis at point **W**.
C Light limits the rate of photosynthesis at point **Y**.
D Light limits the rate of photosynthesis at point **Z**. ( )

10. The diagram shows a section through a leaf with four layers of cells labelled.

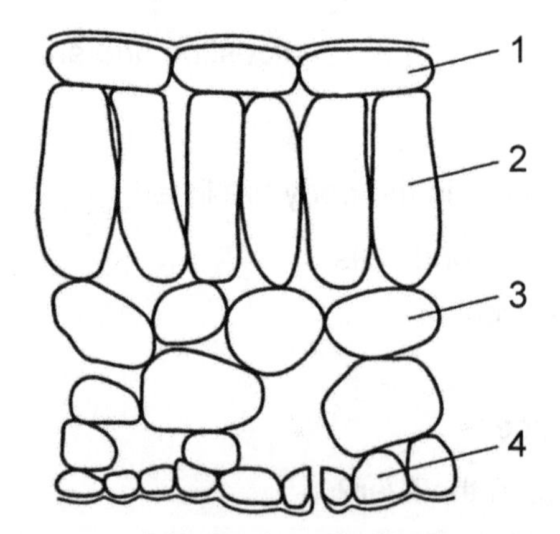

Which row shows the correct comparison between a layer where the cells contain more chloroplasts and a layer where the cells contain fewer chloroplasts?

	layer where the cells contain more chloroplasts	layer where the cells contain fewer chloroplasts
A	1	2
B	1	4
C	2	3
D	4	3

( )

11. When young leaves are being formed on a plant, large quantities of mineral ions are needed.

Where and when is the movement of mineral ions in the plant the greatest?

A companion cells on a hot sunny day
B root hair cells on a cool cloudy day
C sieve tube elements during a warm night
D xylem vessels on a warm sunny day ( )

12. Which graph shows the effect of humidity on the rate of transpiration of a well-watered plant?

A

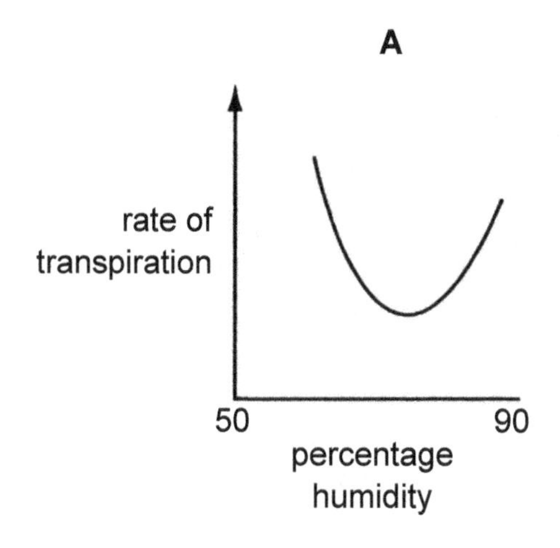

B

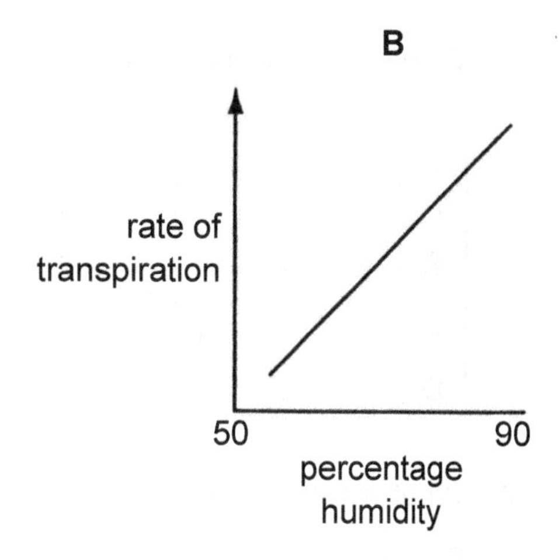

C

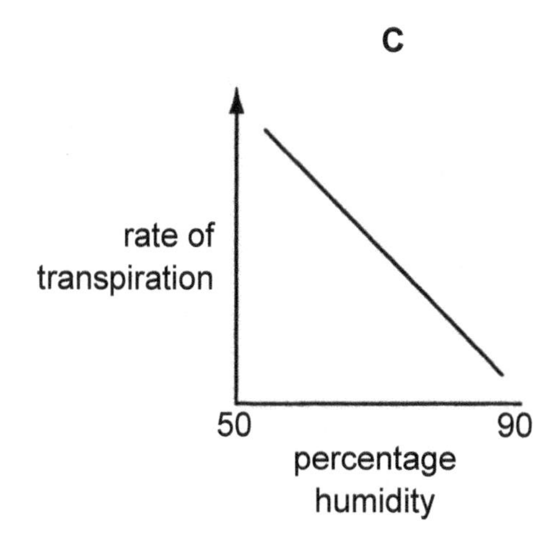

D

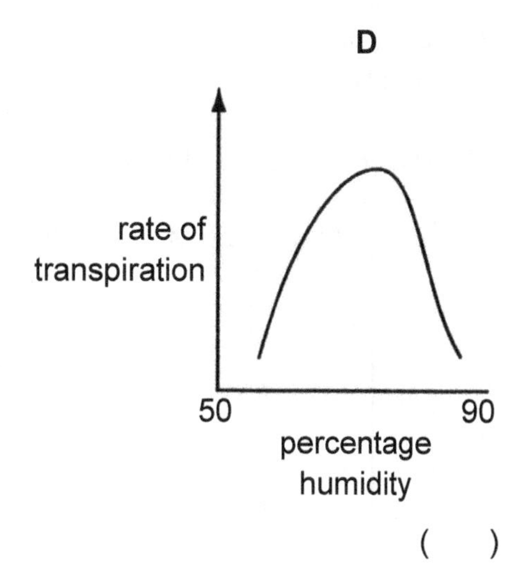

( )

13. The diagrams show sections through a leaf and a stem.

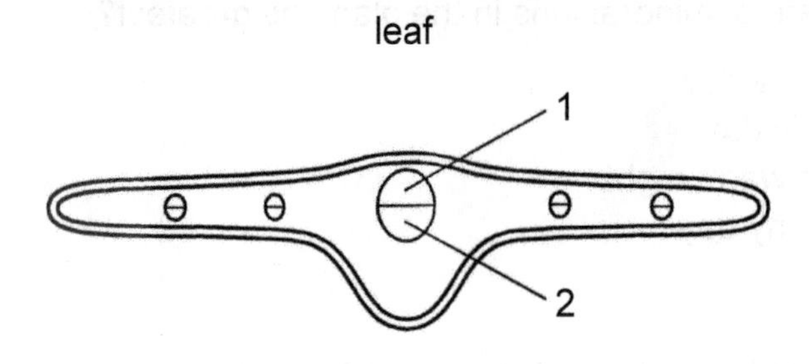

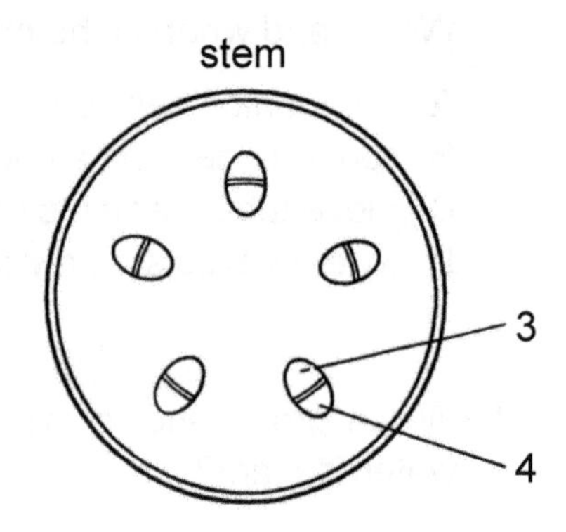

Where does translocation take place?

	leaf	stem
A	1	3
B	1	4
C	2	3
D	2	4

( )

14. What is transported in solution in the blood plasma of a healthy person?

A carbonic anhydrase
B fibrin
C glucagon
D glycogen

( )

15. Which blood transfusion can be carried out safely?

	blood type of donor	blood type of recipient
A	A	O
B	AB	A
C	B	A
D	O	B

( )

16. The diagram shows a group of alveoli and their blood supply.

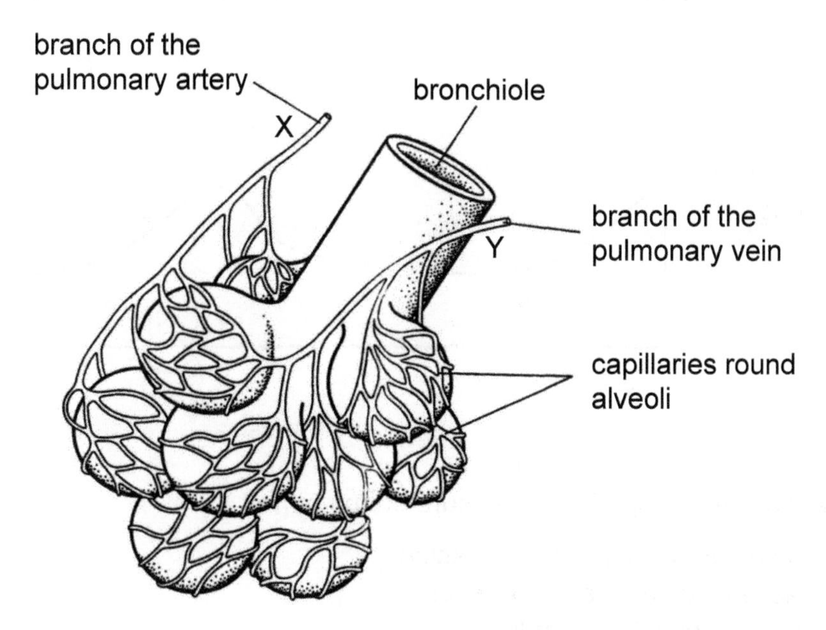

What describes the difference in composition between the blood at X and the blood at Y?

	concentration of chemical in the blood at X compared to the blood at Y		
	carbon dioxide	glucose	oxygen
A	less	more	less
B	less	the same	more
C	more	more	less
D	more	the same	more

( )

17. The graph shows changes in air pressure in the lungs during breathing.

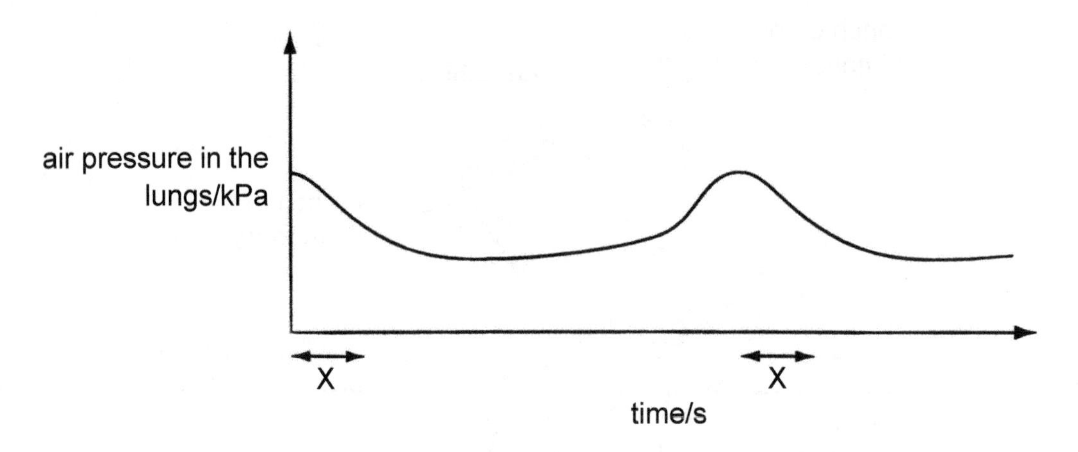

What causes the change in air pressure during period X?

A contraction of the diaphragm muscles
B decrease in the volume of the lungs
C movement of ribs downwards
D relaxation of the external intercostal muscles ( )

18. What are the products of aerobic and anaerobic respiration in humans?

	aerobic respiration			anaerobic respiration		
	carbon dioxide	lactic acid	water	carbon dioxide	lactic acid	water
A	✓	✓	×	✓	✓	×
B	✓	×	✓	✓	✓	×
C	✓	×	✓	×	✓	×
D	×	✓	×	✓	×	✓

key
✓ = is a product
× = is not a product ( )

19. The diagram shows some of the reactions which occur in a red blood cell.

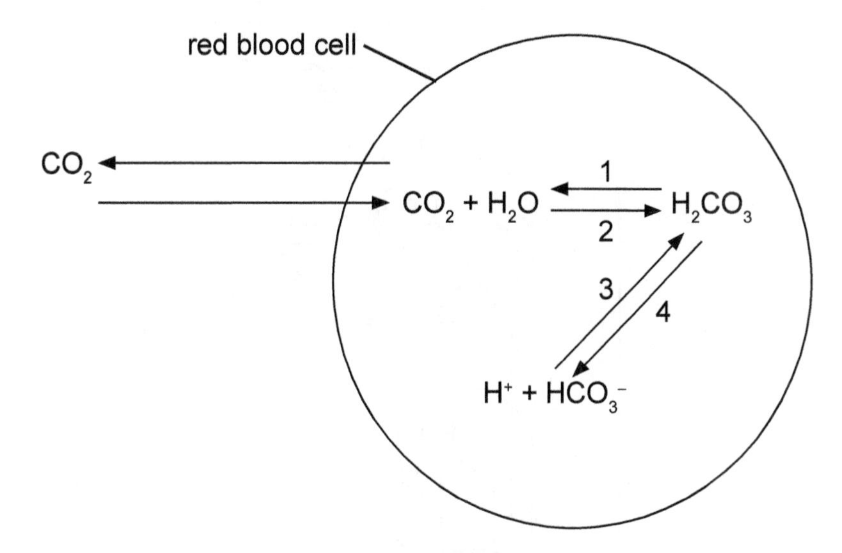

Which numbered reaction(s) involve carbonic anhydrase?

A 1 and 2 **B** 1 only **C** 3 and 4 **D** 4 only ( )

20. Where is the waste product of amino acid metabolism produced and which organ excretes it?

	produced by	excreted by
A	kidney	liver
B	kidney	skin
C	liver	kidney
D	liver	large intestine

( )

21. The diagram represents a kidney tubule and associated blood vessels.

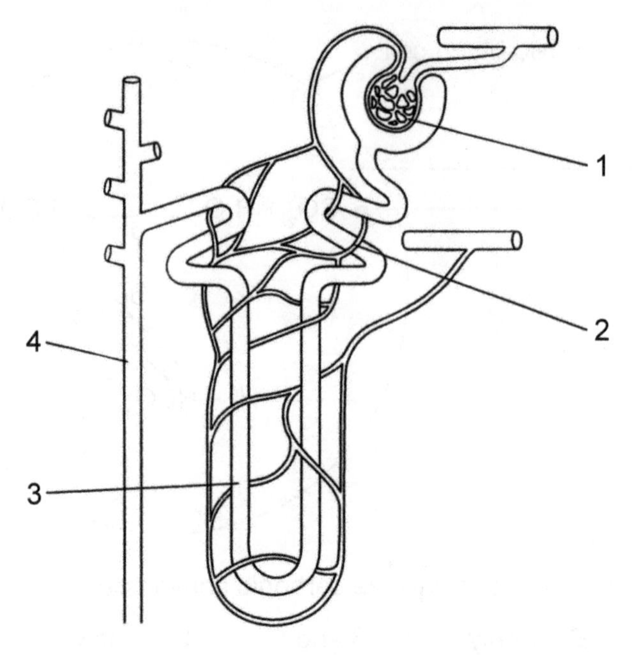

Where does anti-diuretic hormone (ADH) have its effect and where does ultra-filtration occur?

	anti-diuretic hormone	ultra-filtration
A	1	2
B	2	3
C	4	1
D	4	3

( )

22. A swimmer stays too long in very cold water and his body temperature falls below 37°C.

After he comes out of the water, what will help his body temperature return to normal?

1 blood rushing to the skin surface
2 drying the skin quickly with a towel
3 hair erector muscles relaxing
4 shivering
5 running around

A 1, 2 and 3 **B** 1, 3 and 4 **C** 2, 4 and 5 **D** 3, 4 and 5 ( )

23. Which factors are controlled by homeostasis?

	glucose concentration in the blood	water content in the ileum	temperature in the liver
A	✓	✓	×
B	✓	×	×
C	✓	×	✓
D	×	✓	✓

key
✓ = controlled by homeostasis
× = not controlled by homeostasis ( )

24. Which row is correct for adrenaline?

	role of adrenaline in liver	adrenaline broken down in
A	conversion of glucose to glycogen	kidney
B	conversion of glucose to glycogen	liver
C	conversion of glycogen to glucose	kidney
D	conversion of glycogen to glucose	liver

( )

25. When the eye adapts to view a distant object, what are the receptors and effectors?

	receptors	effectors
A	ciliary muscles	iris muscles
B	iris muscles	retinal cells
C	retinal cells	ciliary muscles
D	retinal cells	iris muscles

( )

26. The flow diagram shows the pupil reflex.

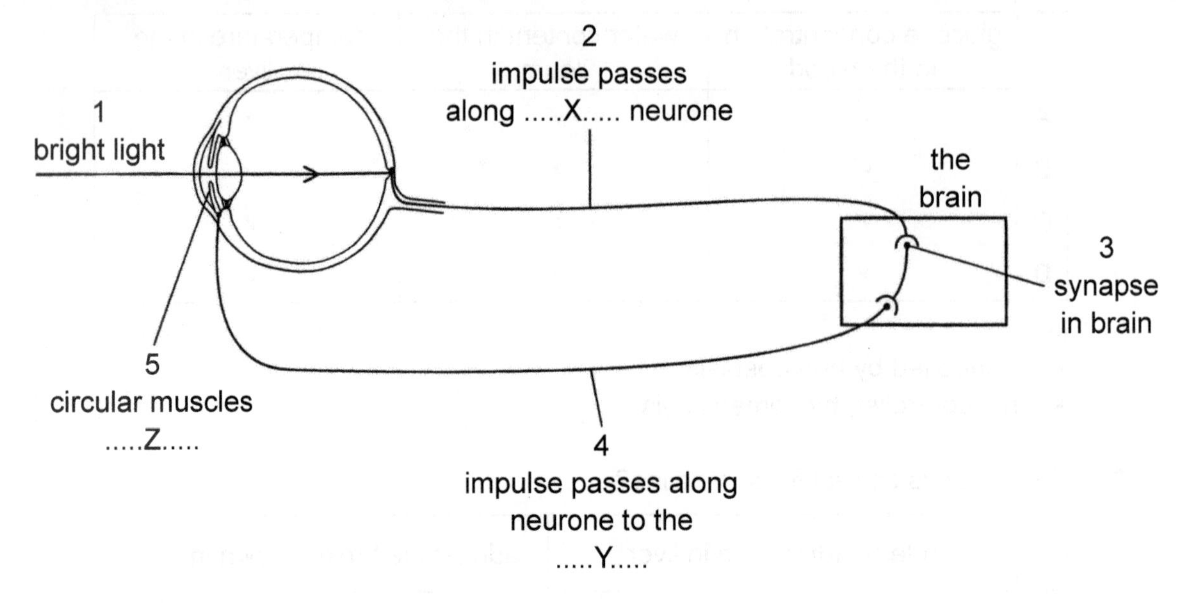

Which words complete the flow diagram?

	X	Y	Z
A	motor	ciliary body	contract
B	motor	iris	relax
C	sensory	ciliary body	relax
D	sensory	iris	contract

( )

27. What is a function of the amniotic sac and its fluid?

A allows more than one zygote to implant in the uterus wall
B lets the fetus move around during development
C protects the lining of the uterus from damage by the fetus
D provides the fetus with nutrients from the mother ( )

28. The diagram shows a section through a flower.

Which labelled part contains cells with haploid nuclei?

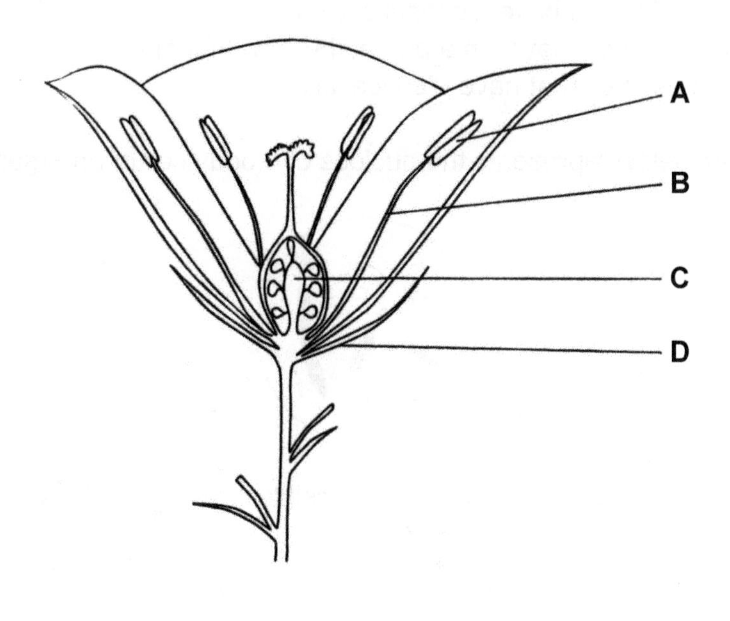

( )

29. During the fertile of a menstrual cycle, which changes are occurring to the uterus wall and to the concentrations of estrogen and progesterone in the blood?

	uterus wall	concentration of estrogen	concentration of progesterone
A	breaking down	falling	falling
B	breaking down	rising	rising
C	thickening	falling	rising
D	thickening	rising	falling

( )

30. Which row gives the correct comparison of blood passing from the fetus to the placenta compared to the composition of the blood flowing from the placenta to the fetus?

	blood from fetus to placenta			
	amino acids	oxygen	urea	water
A	less	less	more	more
B	less	same	same	same
C	more	less	more	less
D	same	same	less	more

( )

31. What describes homologous chromosomes?

A two chromatids that are joined together to form one chromosome
B two chromatids that have identical alleles
C two chromosomes that form a pair at the start of meiosis
D two chromosomes that have identical alleles ( )

32. The diagram below represents the nucleus of a body cell of an organism.

Which diagram does **not** represent a possible gamete nucleus produced by the organism?

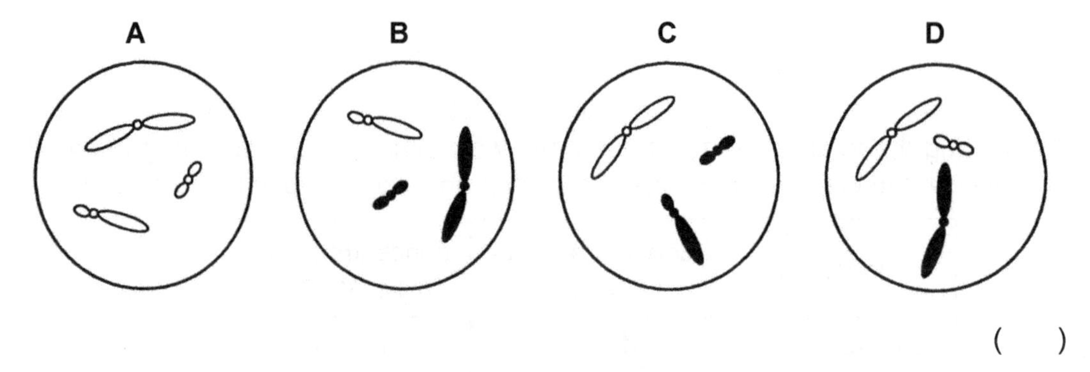

( )

33. The diagram shows some of the events that happen during mitosis.

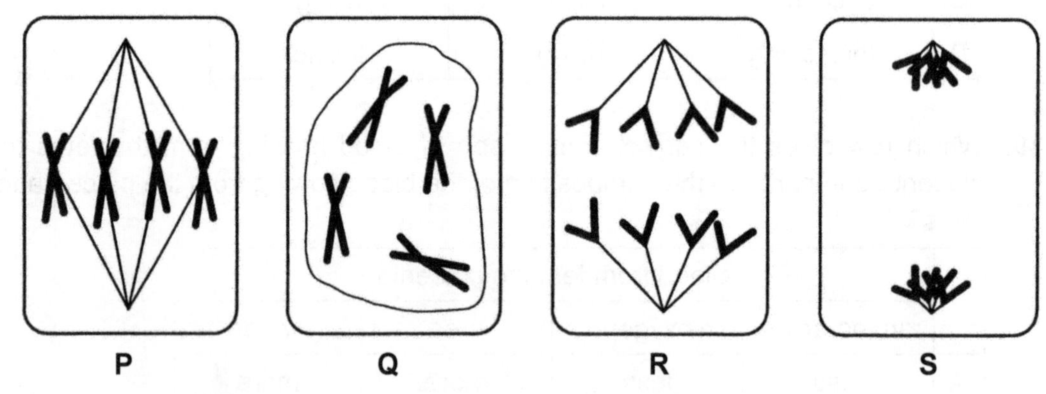

What is the correct sequence for these events during mitosis?

A P → Q → R → S
B P → R → Q → S
C Q → P → R → S
D S → P → Q → R ( )

34. The diagram represents a short section of DNA.

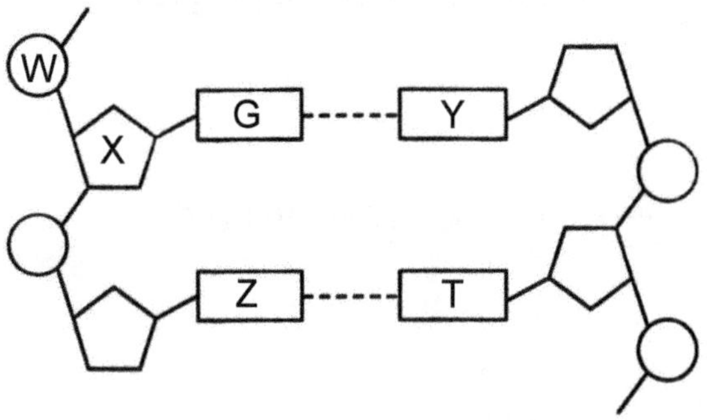

key
G = guanine
T = thymine

Which row correctly identifies the parts labelled W, X, Y and Z?

	W	X	Y	Z
A	adenine	cytosine	sugar	phosphate
B	cytosine	adenine	phosphate	sugar
C	phosphate	sugar	cytosine	adenine
D	sugar	phosphate	adenine	cytosine

( )

35. A human insulin gene can be cut out of human DNA and inserted into the plasmid of a bacterium. The diagrams show four stages of this process.

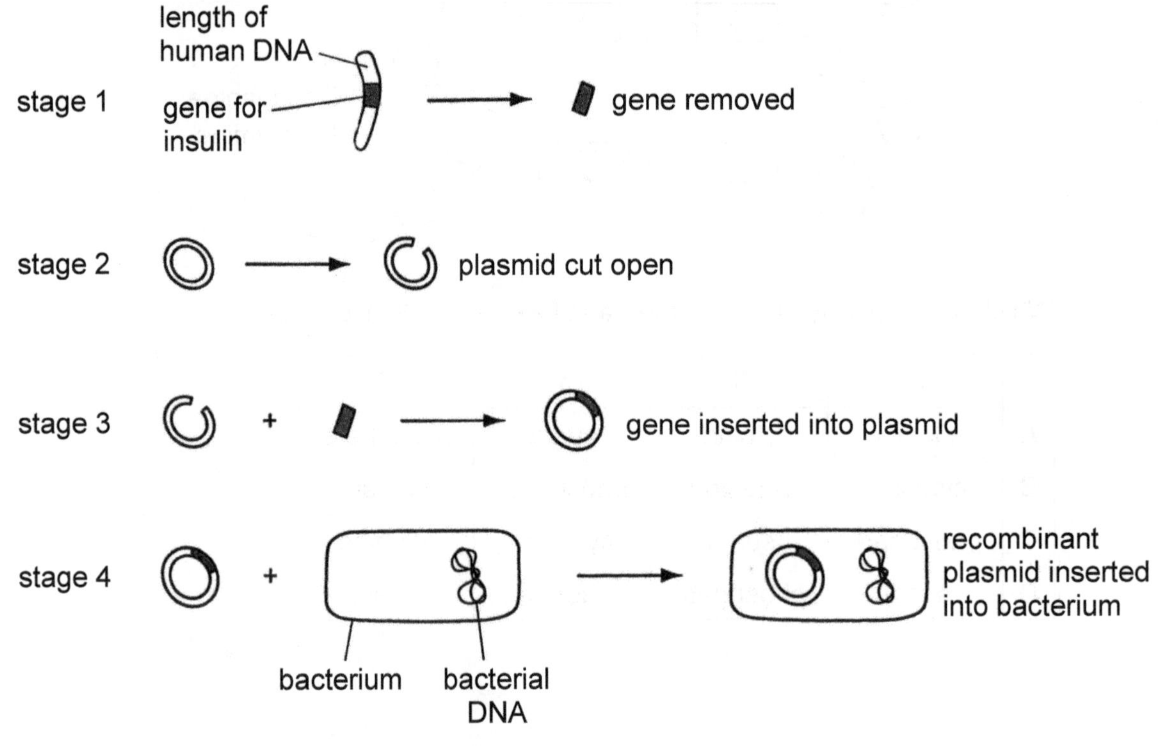

Which row correctly identifies the enzyme used at one of these stages?

	stage	ligase enzyme	restriction enzyme
A	1	✓	×
B	2	×	✓
C	3	×	✓
D	4	✓	×

key
✓ = enzyme used
× = enzyme not used

( )

36. In humans, what describes the chromosomes in the gametes that fuse at fertilisation and result in a normal male zygote?

	chromosomes in the egg	chromosomes in the sperm
A	22 + 1X	22 + 1Y
B	22 + 1Y	22 + 1X
C	23 + 1X	23 + 1Y
D	23 + 1Y	23 + 1X

( )

37. A brown-eyed man and a brown-eyed woman have five children, three with brown eyes and two with blue eyes.

Which statement about this family is correct?

A Both blue-eyed children have the same genotype and phenotype.
B The alleles for brown eyes and blue eyes are co-dominant.
C The genotypes of the parents must be the same as the genotypes of the three brown-eyed children.
D The three brown-eyed children must be heterozygous for eye colour. ( )

38. Which statement uses genetic terminology and information correctly?

A Recessive alleles are usually harmful and dominant alleles are beneficial.
B When two animals are crossed, one homozygous dominant for fur colour and the other homozygous recessive, their offspring will each have different phenotypes.
C When two people, both homozygous for blood group A, have children, both parents and offspring will have the same genotype.
D Under the ABO blood grouping system, alleles I^A, I^B and I^O are co-dominant.
( )

39. In sewage disposal, what are the results of processes that involve microorganisms and of processes that involve anaerobic microorganisms?

	results of aerobic process	results of anaerobic process
A	carbon dioxide and water are produced	methane is produced
B	fats and oils are metabolised	urea is converted to nitrates
C	inorganic solids are removed	methane is produced
D	pesticides are deactivated	ammonia is removed

( )

40. Insect larvae feed on the leaves of trees.

A hawk eats small birds which feed on these larvae.

The diagram shows the pyramid of numbers and the pyramid of biomass for this food chain. Four levels have been labelled A, B, C and D.

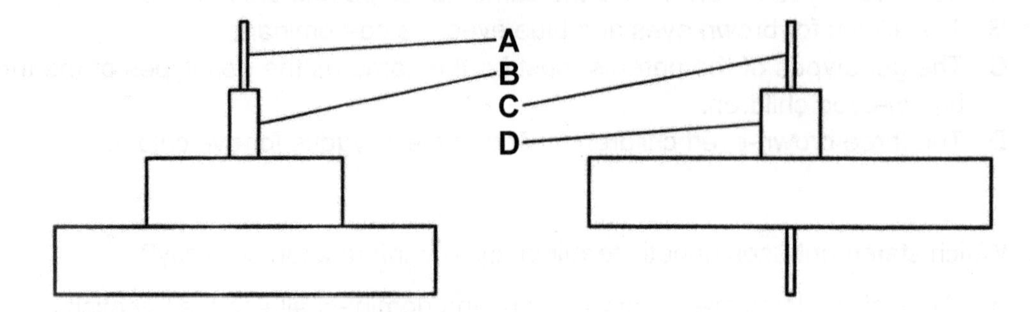

Which level is described correctly?

A biomass of the primary consumers

B biomass of the secondary consumers

C number of producers

D number of herbivores ( )

BIOLOGY

Paper 2

5158/02

October/November 2014

1 hour 45 minutes

Section A

Answer **all** questions.

Write your answers in the spaces provided.

1. **(a)** Define the term *excretion* and explain its importance.

....

....

....

.... [2]

(b) In a survey of 3000 people the minimum and maximum concentrations of certain substances in the urine were determined.
Table 1.1 shows the results of the survey.

Table 1.1

substance	concentration / mg per dm^3		
	minimum concentration	maximum concentration	difference in concentration
urea	9300	23 300	14 000
sodium	1170	4390	3220
chloride	1870	8400	6530
ammonia	17	200	

(i) Complete Table 1.1 to show the difference in concentration for ammonia. [1]

(ii) State **one** other excretory product not shown in Table 1.1.

.... [1]

(c) Describe and explain how reduced secretion of ADH would affect the composition of urine.

....

....

....

....

....

....

....

....[3]

[Total: 7]

2. The quantity of pure alcohol in a drink can be expressed as alcohol units.
One alcohol unit equals 10 cm^3 pure alcohol which is the amount of alcohol the average adult can break down in one hour.
This means that one hour after drinking one unit of alcohol there should be little or no alcohol left in the blood of an adult.

(a) Name the organ which breaks down the alcohol.

....[1]

(b) The strength of an alcoholic drink can be indicated by the percentage of pure alcohol it contains.

(i) Calculate the number of units of alcohol consumed by a person who drank 3×175 cm^3 glasses of wine with a strength of 12.0%.
Show your working.

.... units [2]

(ii) State how long it would take for the body to break down this amount of alcohol.

.... hours [1]

(c) State two short-term and two long-term effects of excessive alcohol consumption.

short-term

1. ....

....

....

2. ....

....

....

long-term

1. ....

....

....

2. ....

....

....

[4]

[Total: 8]

3. Fig. 3.1 shows an X-ray of a *Coleus* leaf.

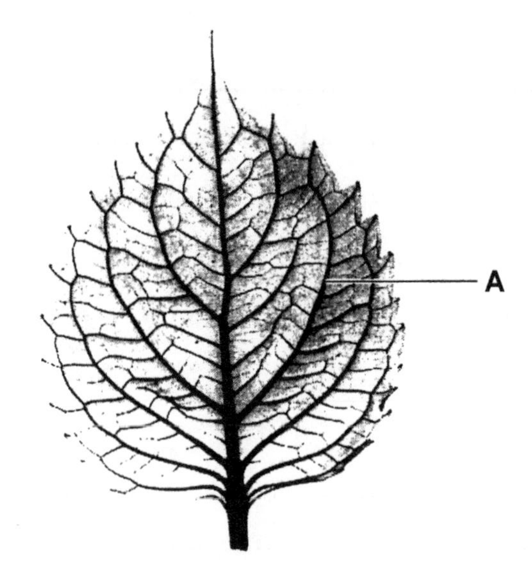

Fig. 3.1

(a) (i) Name the structure labelled **A**.

....[1]

(ii) Name **one** tissue present in the structure labelled **A** and state how it is adapted to its function.

name ....

adaptation ....

....[2]

(b) Fig. 3.2 shows the changes in the rate of water loss from a plant over twenty four hours.

rate of water loss /cm^3 per hour

300
250
200
150
100
60
0

0 2 4 6 8 10 12 14 16 18 20 22 24

Fig. 3.2

(i) Name the process responsible for water loss in a plant.

....[1]

(ii) State the length of time the rate of water loss was greater than 150 cm^3 per hour.

....[1]

(iii) State the **two** times when the rate of water loss was 70 cm^3 per hour.

.... and ....[1]

(iv) State **one** factor, other than temperature, which may be responsible for the changes in the rate of water loss.

....[1]

[Total: 7]

4. **(a)** State what is meant by the term *carbon sink*.

....

....

....

....[2]

(b) Fig. 4.1 shows part of the carbon cycle.
The numbers represent the changes in the flow of carbon in gigatons of carbon per year between parts of the cycle.

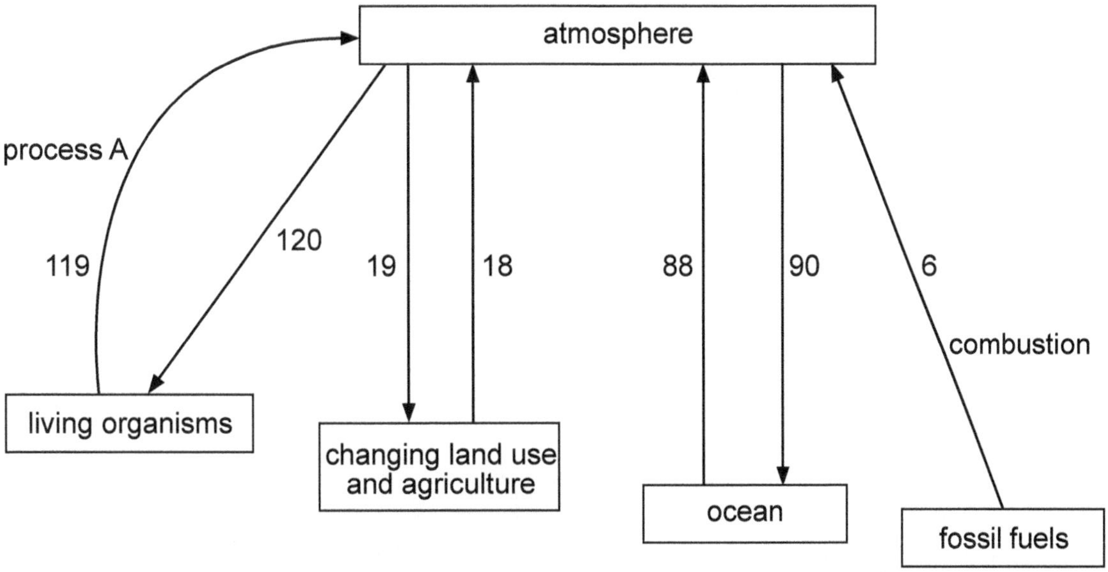

atmosphere
process A
119
120
19
18
88
90
6
combustion
living organisms
changing land use and agriculture
ocean
fossil fuels

Fig. 4.1

(i) Name process **A**.

....[1]

(ii) State the effect that changing land use and agriculture has on the quantity of carbon dioxide in the atmosphere.

....

....[1]

(iii) Calculate the total change in the amount of carbon in the atmosphere each year.
Show your working.

.... [2]

(iv) Suggest why the ocean gives out less carbon than it takes in.

....

....

....

....

....

....

....

....[5]

[Total: 11]

5. **(a)** Explain the effect of carbon monoxide on the normal functioning of the body.

....

....

....

....[2]

(b) Table 5.1 shows results of a study into the number of deaths from two diseases for cigarette smokers and non-smokers.

Table 5.1

cause of death	number of deaths			
	total	smokers	non-smokers	ratio of deaths between smokers and non-smokers
coronary heart disease	3360	1920	1440	4 : 3
lung cancer	360	324	36	

(i) Complete Table 5.1 by calculating the ratio of deaths between smokers and non-smokers for lung cancer.
Show your working.

[2]

(ii) Suggest why the number of deaths from lung cancer in smokers is much greater than it is in non-smokers.

....

....[1]

[Total: 5]

6. **(a)** State what is meant by the term *asexual reproduction*.

....

....

....

....[2]

(b) Fig. 6.1 shows four stages, **A**, **B**, **C** and **D**, in mitosis in bluebell, *Hyacinthoides non-scripta*.

A

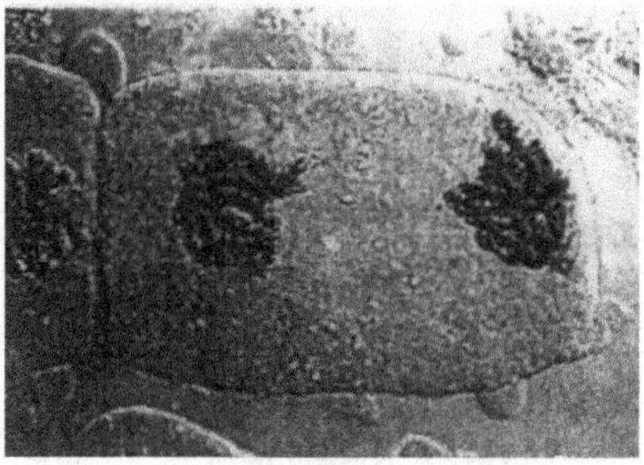

B

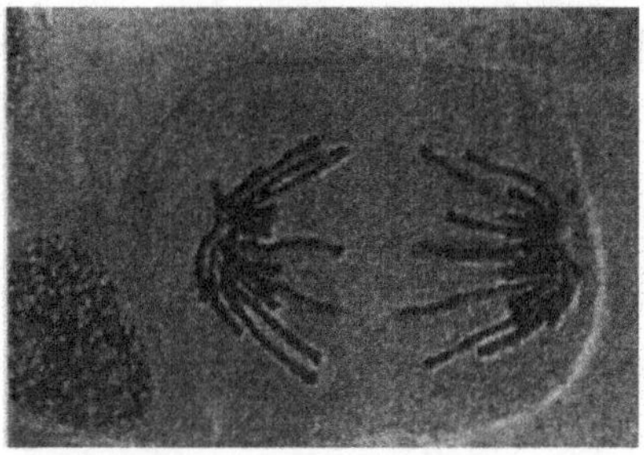

C

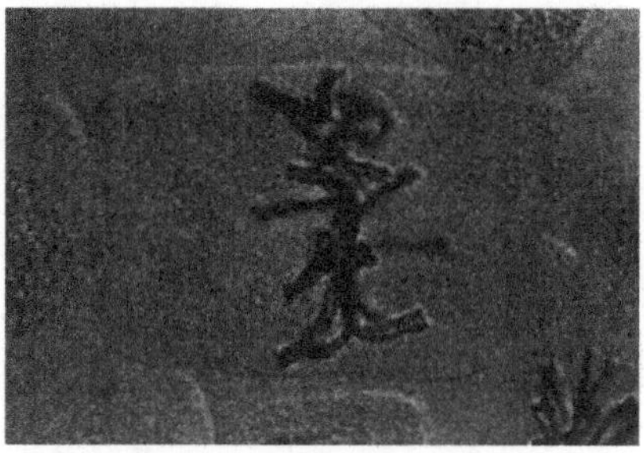

D

Fig. 6.1

Name each of the stages shown in Fig. 6.1.

A ....

B ....

C ....

D ....

[4]

[Total: 6]

7. **(a)** State the three alleles of the gene that controls human blood groups.

1. ....

2. ....

3. ....[1]

(b) Complete the genetic diagram below to show the possible blood groups of the children of a father who is heterozygous for blood group A and a mother who is heterozygous for blood group B.

		father		mother	
genotypes of parents		....		....	
gametes	....		....	....	....
genotypes of offspring	....		....	....	....
phenotypes of parents	....		....	....	....

[5]

[Total: 6]

Section B

Answer **three** questions.

Question 10 is in the form of an **Either/Or question**.
Only one part should be answered.

8. **(a)** State two ways in which the structure of arteries differs from the structure of veins.

1. ....

....

2. ....

....[2]

(b) Fig. 8.1 shows the speed of blood flow as it passes through the arteries, capillaries and veins.

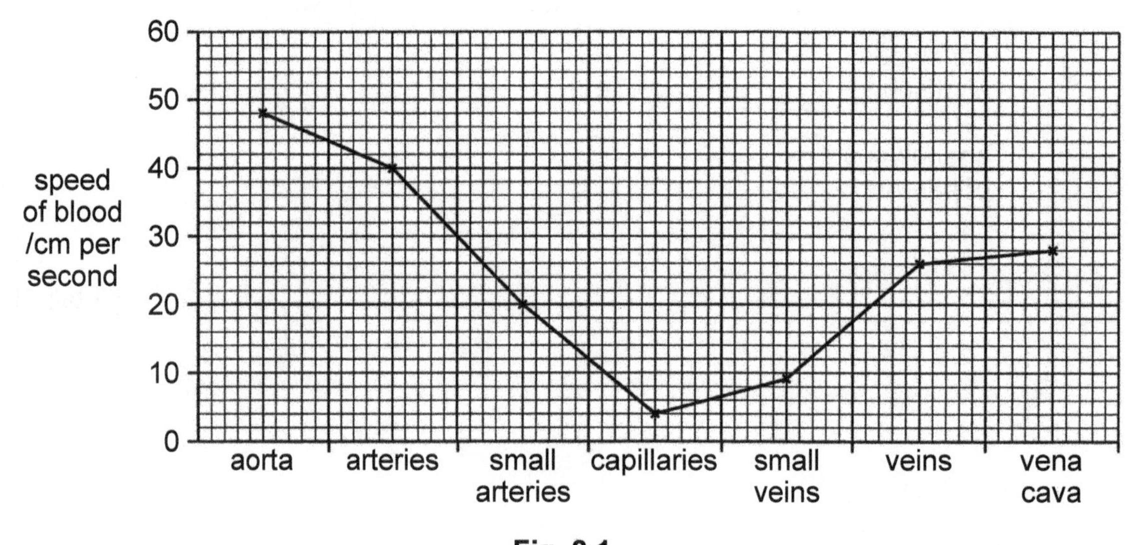

Fig. 8.1

(i) With reference to Fig. 8.1, describe the changes in the speed of the blood as it passes from the aorta to the vena cava.

....

....

....

....

....

....[3]

(ii) Table 8.1 shows the changes in blood pressure as blood passes from the aorta to the vena cava.

Table 8.1

blood vessel	mean blood pressure / mm mercury
aorta	104
arteries	88
small arteries	32
capillaries	18
small veins	12
veins	6
vena cava	4

Plot the data from Table 8.1 onto the grid below. [4]

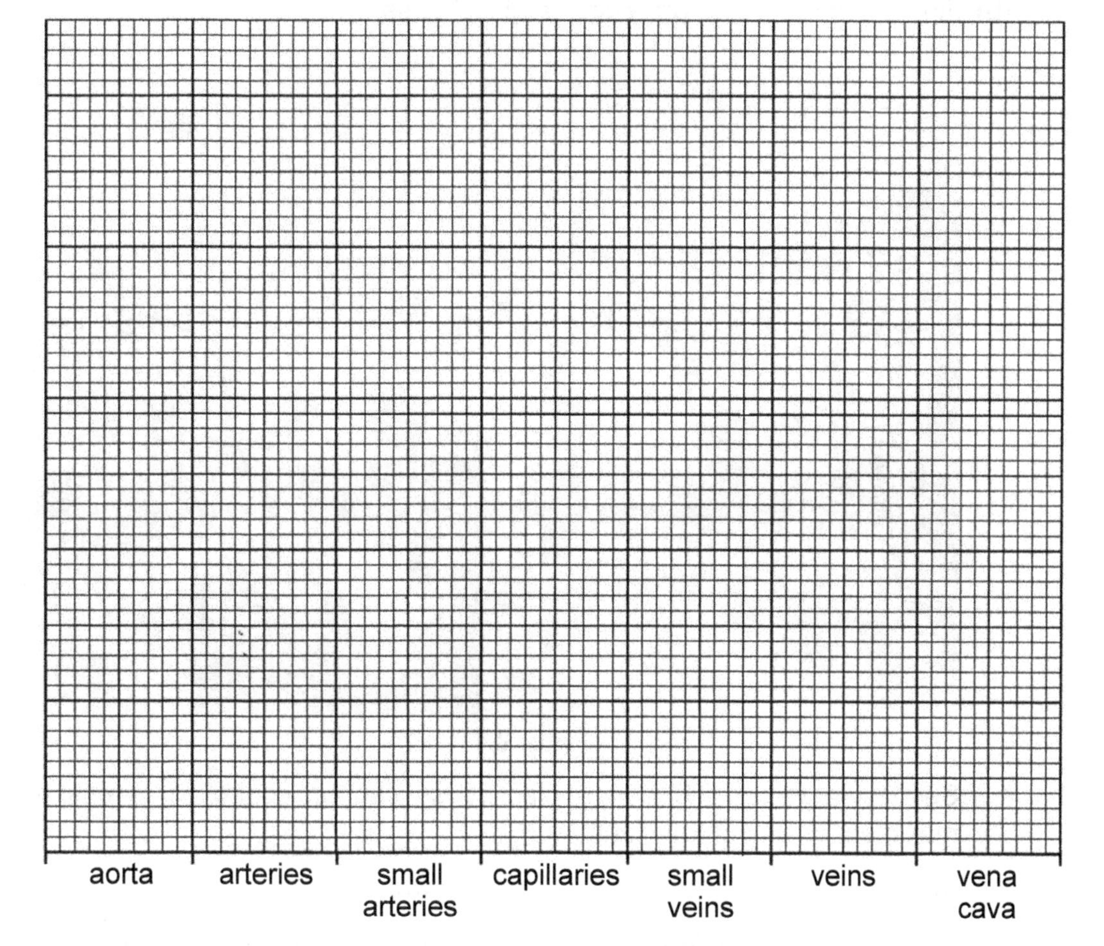

(iii) Describe the process that generates the pressure in the aorta.

....

....[1]

(iv) Suggest what causes the fall in blood pressure as the blood passes from the aorta to the vena cava.

....

....[1]

[Total: 11]

9. (a) Describe how light energy is converted to chemical energy and stored in carbohydrates in plants.

....

....

....

....[4]

(b) Fig. 9.1 shows plants growing in a glasshouse.

Fig. 9.1

Suggest how conditions in a glasshouse can be controlled to ensure the maximum growth of the plants.

....

....

....

....

....[5]

[Total: 9]

10. EITHER

(a) Explain how air is made to enter the lungs.

....

....

....

....

....

....

....

....

....

....

....

....

....

....[6]

(b) Describe how a molecule of oxygen present in the air breathed in reaches a muscle cell in the wall of the left atrium.

....

....

....

....

....

....

....

....

....

....[4]

[Total: 10]

10. Or

(a) Describe the properties of enzymes.

....

....

....

....

....

....

....

....

....

....

....

....[6]

(b) Biological washing powders (detergents) contain one or more enzymes.

Suggest the advantages of using biological washing powders compared to those without enzymes.

....

....

....

....

....

....

....

....

....

....[4]

[Total: 10]

NOTES

NOTES